Contents

Page

LIVESTOCK ETHICS

Respect, and our duty of care for farm animals

By

Rev Dr Gordon J Gatward

Director, The Arthur Rank Centre, National Agricultural Centre, Stoneleigh Park, Warwickshire, CV82LZ, United Kingdom

CHALCOMBE PUBLICATIONS

First published in Great Britain by
Chalcombe Publications
Painshall, Church Lane, Welton,
Lincoln, LN2 3LT.
United Kingdom

© G.J Gatward 2001

ISBN 0 948617 46 2

Foreword

By Mike Calvert, Chief Executive, Royal Agricultural Society of England

There have been few occasions in the past century when livestock farmers have felt more isolated, more helpless, more insecure than now. As I write, the foot and mouth crisis in the United Kingdom is still unfolding. The Royal Agricultural Society of England and The Arthur Rank Centre are inundated with people seeking guidance as they attempt to cope with the impact of the disease on their lives.

Gordon Gatward is at the centre of the livestock farmers' situation. As a sheep breeder, stockman, agricultural chaplain and Director of the Arthur Rank Centre at the National Agricultural Centre, Gordon knows better than most people how the lot of the livestock farmer has changed over the years, as farms have increased in size and intensity of operation. He understands the essence of the animal and identifies with the livestock keeper. The central theme of this timely book is that the ethical basis of livestock production is respect for the essence of the animal, expressed in skilled stockmanship and responsible stewardship.

Although originally an academic thesis, this book is as readable as any other on livestock husbandry. Gordon Gatward has kept his practical approach, even when writing about philosophical matters like the human/animal relationship. As a result, many parts are of direct help to those farmers and livestock keepers who are seeking support and guidance in their routine daily work with animals. I urge those of you who are not directly involved with livestock and livestock keeping to read this book, because it will make you think about the central issues involved in the industry. It will also clarify many misconceptions about livestock farming. This book is far from a defence of present practice; it is a clarion call to all who care for livestock to consider what is acceptable and what is unacceptable, and then to take appropriate action to ensure that the future of our livestock industry is grounded in the sound ethical base spelt out in these pages.

Preface

Over the years my life has been enhanced and enriched by the company of countless stockmen. From them I've not only learnt many practical skills but have discovered a great deal about the relationship that can exist between humans and animals and the responsibility a farmer has for his stock. In many cases they also provided me with the bedrock of much of my theology and faith. This work is a special tribute to them, and in particular to Willie Airey, Norman Chapple, Thomas Raw, Aubrey Wiseman, and especially my own father-in-law Arthur Blake, whose skill with stock I will always try to emulate but know I am unlikely to achieve.

Both the research, and this finished product, also my PhD thesis, owe everything to the patience, encouragement, guidance and criticism of my supervisors at the University of Leeds, Jennifer Jackson and Mike Wilkinson, to whom I am more than grateful. Despite close personal bereavement and time spent working in Canada, Jennifer was always accessible and ready to offer guidance and advice. I am especially grateful to her for the depth of her questions and for her constant encouragement. Mike is really responsible for initiating this work in the first place in that he responded to an initial enquiry as to what I might do with a three month sabbatical by suggesting the study that led to the thesis. His friendship, understanding and support have been greatly appreciated, especially as he has had to cope with major family illness. When I first approached him I never thought that a three-month period of sabbatical leave from my post as Agricultural Chaplain in Lincolnshire would become five years' study.

The Yorkshire Agricultural Society provided me with a grant at the start of my studies to help meet the cost of travelling regularly to the University of Leeds. I am grateful for their assistance, which has been much appreciated.

Above all my thanks go to my family for their patience and perseverance, especially to my wife Janet. Countless evenings, weekends and holidays have been absorbed by my concentrating on this work instead of on home and family responsibilities. Janet has suffered it all and still managed to smile whilst delivering countless cups of coffee and saying all the encouraging things I needed to hear.

My thanks go to them all and my hope is that this piece of work proves a fitting tribute to everyone who has helped and supported me so much over the past five years and during a lifetime working with and enjoying the company of livestock and their keepers.

Gordon Gatward
The Arthur Rank Centre
Stoneleigh
Warwickshire
CV8 2LZ

February 2001

viii

CHAPTER 1

Introduction

Since the end of the Second World War, the emphasis in British agriculture, as in most developed nations, has been on ever-greater efficiency and productivity at ever-decreasing unit cost of production. Epitomised in the UK in both the 1947 Agricultural Act (HMSO, 1947), and the Treaty of Rome (1957), both of which are discussed in Chapter 2, this outlook has encouraged the development of intensive systems of livestock production, defined in Chapter 2, which have been widely criticised for their apparent exploitation of both livestock and the environment.

These criticisms, when applied to intensive methods of livestock production, are largely focused on the impact that such systems may have on the welfare of the animals concerned and the ethical issues that are consequently raised. The Royal Society for the Prevention of Cruelty to Animals (RSPCA), for example, has expressed its opposition to the use of intensive methods and have stated that such systems have been 'designed around the needs of man and have ignored the welfare of their occupants' (RSPCA, 1995:3). Likewise, The Farm and Food Society has expressed its misgivings at the ethical implications of intensive production systems: 'The conception that man should be provided with over-abundance, obtained by any and every means of exploiting animals, plants and soil, is of this age. Man's true instinct has given way to a strange reasoning, which puts ever-increasing production as the goal, without account of need, environment, health, humanity or effects on future generations.' (The Farm and Food Society, 1972).

In this book I examine such wide-ranging generalisations and challenge many of their implicit assumptions. Not the least of these is the criticism that 'ever-increasing production' necessitates the ignoring of ethical issues, and that intensive systems are, by definition, unethical. In a world with a burgeoning population there is a general

9

belief amongst agriculturalists that they have a moral responsibility to provide the food that that population requires, that they believe is the industry's *raison d'être.* 'Hunger and poverty have in history haunted a land in the absence of farming skill, and still do today in the under-developed countries of the world' (Russell, 1963:17).

Some critics of modern husbandry methods suggest that the hungry could be fed more efficiently if there was less animal production and a far greater emphasis on crop husbandry (Rifkin, 1992:290), but such suggestions tend to ignore the environmental and climatic restrictions on crop production in many areas, the culture and traditional skills of the people concerned, and the subsequent dependence of communities on the farming of livestock.

Another assumption often made by opponents of intensive agricultural production systems is that both their own position and their arguments are morally unassailable. I believe that this also needs to be challenged. Singer for example (1990:xxiv) has stated that 'The strength of the case for Animal Liberation is its ethical commitment; we occupy the high moral ground.' I suggest that this is far from being the case. Indeed, as I indicate in Chapter 2, a number of philosophers now seriously question the validity of Singer's claim, and argue that there is an ethical case for continued livestock production and even the applications of certain intensive methods. This book develops that argument, in the belief that the future of livestock production depends on a valid case being made in its favour. Animal rights activists have pledged themselves to end what they see as the exploitation of animals (Regan, 1984:351). If the British livestock industry is to survive, then a valid case must be made for the ethical acceptability of intensive production systems. I aim to establish that the harmonisation of ethically acceptable standards of animal welfare with systems of intensive livestock production can be achieved.

On a cold winter Friday in 1998 a group of angry men gathered round one of the sheep pens at Louth auction market in Lincolnshire. In the pen was an elderly Suffolk ewe. She was emaciated and had a number of sores on her back. She could hardly stand let alone walk. To make matters worse her efforts at locomotion were impeded by the grossly overgrown horn of her feet. That sheep, whilst not being the subject

10

of the men's anger, was certainly the cause. The person who had delivered her to that pen to be sold, and then quickly disappeared, was cursed and criticised. His efforts at shepherding were loudly condemned, and his safety if he had been present would have been questionable. Those men were offended that an animal should have been treated as that sheep had been. Its condition and suffering contradicted all that they believed in and stood for as stockmen and this was evidenced in their overwhelming expression of compassion for that creature.

Whilst words such as 'cruel', 'wicked' and 'criminal' were being voiced I wonder how many of those who stood by the sheep's pen could have given a reason for the strength of their feelings or could have accounted for the apparent contradiction between their sensibilities and their profession. Many of them were producing livestock for the meat trade in conditions that some would be quick to criticise and yet those same men were angry and offended at the way one elderly ewe had been treated. If challenged how many of those men would have been able to give a reason for this apparent paradox? Yet their understanding of humane feelings for their stock and their implicit and inborn sense of responsibility for the animals in their care is fundamental to any debate concerning either farm animal welfare or the ethics of livestock production systems.

In the New Testament, Christians are instructed to 'always be prepared to give an answer to everyone who asks you to give the reason for the hope that you have' (I Peter 3:15). When beliefs and principles are under attack those who hold them must be able to articulate the reason for believing what they do, or for living in the manner that they choose. As the following pages will show, livestock production, especially for slaughter, is in that situation. The purpose of this book is to argue the case of the livestock producer, to give voice to those implicit principles and deep sense of responsibility which the vast majority of stock keepers feel but probably find it difficult to articulate, and to provide the rationale for the ethical acceptability of some of those husbandry practices and systems that are often misunderstood and criticised.

Furthermore, if livestock producers are to answer their critics they must be able to provide the criteria by which production systems can be ethically evaluated. In this book I hope to provide such an ethical evaluation and to give a reasoned explanation for the anger of those men in Louth mart.

The book includes an analysis of some of the current welfare criticisms of intensive systems and, where these are valid, an examination of the implications of introducing the required changes in practice. It will be grounded in an historical examination of the developments in livestock production, in a critical survey of the arguments propounded by philosophers and theologians on issues related to the livestock industry and in an appreciation of the need for renewed emphasis on the importance of the highest levels of stockmanship expertise, training and skills in intensive systems.

CHAPTER 2

Ethical Issues

2.1 Intensification *versus* extensification.

The options now open to farmers regarding the systems they might employ to produce the raw materials required by the food processing industry are many and varied, with each one increasingly likely to raise questions of an ethical nature, as defined in Section 2.5. Some of the issues that challenge the livestock producer in particular can have a global significance as they range from concerns about the environmental impact of certain systems to direct criticism of livestock production as an inefficient way of meeting the world's food needs and as an irresponsible waste of those agricultural products which are used to feed farm animals (Rifkin, 1992:4).

One of the most voiced concerns with regard to the livestock industry, at least in the developed world, is that many of its methods are considered to be detrimental to animal welfare (Harrison, 1964). Since World War II there has been a move in most western countries towards more intensive systems of farming (defined in Section 2.5), and the adoption of these practices has given rise to many of the current concerns.

The use of battery cages in the poultry industry and tethers and crates by pig producers are obvious examples of practices that have become a familiar focus of the concern and anxiety felt by many members of the public. Although normally hidden from most people's gaze, these systems and the practices they employ have become familiar to the wider public largely as a result of the publicity they have received from groups such as the Royal Society for the Prevention of Cruelty to Animals (RSPCA) and Compassion in World Farming (CIWF). Organisations such as these, dedicated to the promotion and advancement of animal welfare, have made no secret of their opposition to all forms of intensive livestock production, seeing them as cruel and exploitative of animal life. 'Such systems are characterised by an approach whereby a producer's dominant concern is to maximise the yield from the animals concerned, with little attention being given to the impact of the production system on the animal's health and welfare. The animals are regarded simply as units of production to be used as efficiently as possible rather that as sentient beings capable of suffering when pain or ill-health is imposed on them, or when their behaviour or social needs are denied' (Stevenson, 1997:2).

Although the RSPCA appear to have adopted a more balanced approach, they have expressed opposition to many features of intensive systems. Thus they have criticised breeding programmes which in 'attempting to make farm animals grow bigger, faster, or leaner, have already resulted in animals which naturally develop painful abnormalities.' Criticism has also been made of those 'housing systems (which) keep animals in cramped or barren surroundings. These systems have been designed around the needs of man and have ignored the welfare of their occupants' (RSPCA, 1995:3).

Whilst the element of truth in both statements has to be recognised, these comments do not reflect the whole picture. The fact is that there have been many positive benefits to both animals and humans as a result of recent developments in livestock husbandry. Contrary to the suggestion of some groups such as CIWF, there is evidence to prove that high welfare standards can be achieved in intensive systems as long as there is an equal emphasis on high standards of stockmanship. Examined by Varley (1991), Seabrook (1988), and Hemsworth (1997), this evidence will be considered more fully in Chapter 5.

Advances in veterinary medicine, in the understanding and appreciation of an animal's physical and mental needs, in husbandry techniques, in housing, and in agricultural education, are all welfare developments resulting at least in part from the move towards more intensive systems of production. 'The complaint that high-tech farming is bad for animal welfare is only partly true. It is in danger of missing the point that technology has also brought gains for animal welfare and - on some farms - may continue to do so' (Editorial, *The Economist*, 20.4.96:93-95).

In a similar vein, Webster (1991:160-162) compares a badly managed flock of sheep kept under traditional extensive methods, with one that is responsibly managed but shepherded in an intensive system. Whilst the latter is well-housed, well-fed, and in good health the other sheep are 'outwintered and overstocked on a muddy, parasite-infested paddock close to a housing estate. They are malnourished, cold, crippled with foot rot and riddled with worms. They are regularly worried by dogs and some have been killed.'

The point is that intensive systems cannot always be equated with cruelty and extensive practices do not always epitomise kindness. If high standards of welfare are to be achieved in any system of livestock production, then equally high standards of stockmanship and management are required. This is recognised in the Preface to FAWC's 'Codes of Recommendations for the Welfare of Livestock': 'The basic requirements for the welfare of livestock are a husbandry system appropriate to the health and, so far as practicable, the behavioural needs of the animals and a high standard of stockmanship. Stockmanship is a key factor because, no matter how otherwise acceptable a system may be in principle, without competent diligent stockmanship the welfare of the animals cannot be adequately catered for' (Farm Animal Welfare Council, 1991:2).

Besides recognising that there have been welfare benefits directly attributable to the process of agricultural intensification and the developments in agricultural technology, there have been consequential benefits in terms of food production, which also need to be appreciated. 'Increasing numbers of people in rich countries have doubts about modern farming methods. But few, so far, are willing to give up the cheap food these methods have brought. Even fewer consumers in developing countries are willing to make the sacrifice, for farming technology has saved millions of them from starvation .....without modern farming technology more people would be starving today' (Editorial, *The Economist*, 20.4.96:93-95).

The public's concerns regarding the welfare of farm animals has largely focused on this dichotomy between intensive and extensive farming, and the subsequent question as to which is ethically and

15

practically more acceptable. The debate that has ensued and the search for an answer to that question are at the heart of this study.

2.2 The impact of intensive livestock production on the animal welfare debate.

Many of the current anxieties with regard to farm animal welfare are a direct result of the post-war move towards an ever-greater intensification of production. Some sectors of agriculture had adopted more intensive methods long before the Second World War. For example, dairy cattle in Victorian cities were often permanently housed and battery systems for egg production were introduced into the UK from America during the 1920s. The widespread adoption of such methods was largely as a result of three factors: the 1947 Agriculture Act, the increase in population due to improved health, and the post-war boom in technology and mechanisation. The first two provided the motivation, the third the means of achieving it.

The purpose of the 1947 Act was 'to promote a stable and efficient industry capable of producing such part of the nation's food as in the national interest it is desirable to produce in the United Kingdom and to produce it at minimum prices consistent with proper remuneration and living conditions for farmers and workers in agriculture and with an adequate return on capital invested' (cited by Hill, 1975:61). It was this commitment to optimum production at the least possible cost, which set the standard for UK farming from the 1940s. Efficiency and productivity became the twin goals for British agriculture. In fact they retained that position for nearly fifty years until emerging public concerns over such matters as the environmental impact of agro-chemical application undermined their adequacy as long-term objectives for the industry.

History has shown however, that farmers responded wholeheartedly to their task and over the years they achieved the dual objective of increased production and lower prices. In the post-war period costs were reduced to such an extent that whereas 30% of total consumer spending in 1940 was on food, by 1995 this had dropped to just 11% (Stevenson, 1997:4). This success was largely due to the widespread introduction of intensive systems on UK farms. As a result, whereas in 1965 only 9% of egg-laying birds were housed in battery units (Carnell, 1983:3), by 1978 this had risen to 57%, and by 1994 to 85% (Whittemore, 1995).

This move towards increased intensification of production in the poultry sector was reflected in most other areas of the UK's livestock industry with the result that 'real prices of livestock products have declined considerably over the post-war period ... increased levels of agricultural self-sufficiency have been achieved while the population has increased by around a half' (Carnell, 1983:4).

There was growing concern during the 1950s and 1960s, however, that all this was being achieved at the expense of the animals' welfare. Cost-cutting, labour-saving methods linked to increased mechanisation were perceived to be detrimental to the well-being of the livestock. Ruth Harrison's 'Animal Machines', published in 1964, first brought the issues of farm intensification and animal welfare to the attention of the wider public and at the same time first introduced the phrase 'factory farming' into common parlance (Harrison, 1964:1). At the same time it introduced other emotive and now only too familiar concepts into the debate.

This is vividly illustrated in some of Rachel Carson's comments in her foreword to 'Animal Machines' in which she states that 'modern animal husbandry has been swept away by a passion for 'intensivism'; on this tide everything that resembles the methods of an earlier day has been carried away. Gone are the pastoral scenes in which animals wandered through green fields or flocks of chickens scratched contentedly for their food. In their place are factory-like buildings in which animals live out their wretched existences without ever feeling the earth beneath their feet, without knowing sunlight, or experiencing the simple pleasures of grazing for natural food - indeed, so confined or so intolerably overcrowded that movement of any kind is scarcely possible' (Harrison, 1964:vii). The imagery regarding traditional extensive farming methods and the standards of animal welfare that by implication stem from it creates the impression of an idyllic existence. Yet, as has already been indicated, this could often be far from the case. As a consequence I suggest that Carson's comparison between

17

intensive and extensive systems is misleading and motivated as much by emotion as it is by careful observation.

The publication of 'Animal Machines' marked a turning point in the public's perception and understanding of farm animal welfare. Indeed such was its impact that the establishment of a UK Government Committee of Enquiry under the chairmanship of Professor Brambell into 'The Welfare of Animals kept under Intensive Livestock Husbandry Systems' was directly attributable to Harrison's work.

The Committee's report was published in December 1965 (HMSO, 1965) and led in turn to a number of major developments in the field of farm animal welfare. Of these the most effective and far-reaching was probably the appointing of an advisory committee to the government on matters pertaining to animal welfare. In 1979 this committee was re-established as The Farm Animal Welfare Council (FAWC) and it has since become one of the key bodies in this whole area of concern regarding livestock production.

For the farmer, however, the pressure to intensify production as a means of achieving ever-greater efficiency was as marked as ever. What the 1947 Act had encouraged of UK farmers, the Treaty of Rome in 1957 then asked of farmers across Europe, as they were encouraged 'to increase agricultural productivity by promoting technical progress and by ensuring the rational development of agricultural production and the optimum utilisation of all factors of production, in particular labour (Article 39 (1), Brooman and Legge, 1997:190). As the basis of the European Community's agricultural policy, and following the UK's entry into the European Community (EC) this single statement appeared to be a confirmation of the direction taken by British farmers in the post-war years. As Brooman and Legge commented: 'In the light of Article 39 (1) it is not surprising that farmers in Europe received a clear message from the EC to farm as intensively as possible, which in turn led to increased welfare concerns' (Brooman and Legge, 1997:190).

Once again the suggestion is made that intensive systems by their very definition must be detrimental to animal welfare. As has already been indicated, however, the validity of this assumption is open to question, as indeed is the legitimacy of the related assumption that the only alternative is a return to the more extensive systems of the past. I suggest another possibility; and argue that compatibility can be achieved between the efficiency and technology of intensive farming systems on the one hand and the moral demands for the adoption of high welfare and stockmanship standards on the other. To achieve this, however, the public may well have to accept an increased cost in the price of food as their part in supporting the welfare and well-being of farm animals.

This has been recognised by The House of Commons Agriculture Committee in their report 'Animal Welfare in Poultry, Pig and Veal Calf Production 1980-81' (HMSO, 1981). In this it was stated that the Committee did 'not accept the contention, frequently stated or implied, that the public demand for cheap food decrees that the cheapest possible methods of production must be adopted ... society has the duty to see that suffering is not caused to animals, and we cannot accept that duty should be set aside in order that food may be produced more cheaply. Where unacceptable suffering can be eliminated only at extra cost, that cost should be borne or the product foregone.'

There are undoubtedly welfare issues arising out of some intensive practices and these must be resolved. Some of these were highlighted in a report produced by The Soil Association and The Food Commission in which it was stated that as a result of their confinement 30% of battery hens had broken bones when they were slaughtered and 25% of dairy cows had to be culled at a young age because of mastitis or lameness resulting from over-production (Living Earth and The Food Magazine 1995:12-13).

These and similar issues must be addressed and as a result will be discussed in more detail in Chapter 6. As has already been stated I believe that their resolution will not be achieved by returning to out-dated or uneconomic systems, but by arriving at the position described above. Furthermore, I believe that this balance between modern production systems and the dictates of conscience regarding animal welfare requires the establishment of an appropriate ethical principle along with the necessary criteria by which the moral acceptability or otherwise of the systems under question can be assessed. Such is the purpose of this book.

2.3   The emergence of a 'welfare consciousness'

The current widespread public concern for and commitment to the welfare of farm animals was vividly illustrated during 1995 by the large numbers of people from all walks of life who blockaded airports and docks in protest at the export of live animals for slaughter. One reporter noted how the strength of feelings generated by their concern for these animals had caused those unused to demonstrating to take to the streets. He observed that 'although TV cameras focus on the violence of a few extremists, the truth is that 99% of those standing in quiet protest are law-abiding citizens, most of whom have never attended a public demonstration' (Potter, 1995:30).

Although many members of the public are clearly so motivated by their concern for farm animals that they protest in this uncharacteristic way, there is evidence to suggest that many others are still reticent to meet the additional costs that more welfare friendly production systems will entail. This was reflected in the results of a survey commissioned by the National Farmers Union (NFU, 1988). A random sample of 2000 people was interviewed across 136 parliamentary constituencies. In replying to questions about farm animal welfare, although 33% of the respondents expressed concern about the use of battery cages only 55% wanted the use banned. The reason given for opposing a ban was that it would result in additional costs. Indeed, 80% stated that they were unwilling to pay any more for eggs even if they knew that they were produced in alternative systems.

Another survey, again of 2000 people and dealing with similar concerns, was conducted by the National Opinion Poll (NOP) for the RSPCA's Freedom Food scheme in 1997 (RSPCA, 1997). This revealed that there had been some change in attitudes in that 78% now said that they wanted to see better welfare conditions for animals on UK farms. Of these, however, still only 44% admitted to taking welfare into account when buying animal products, although 69% did say that they were willing to pay extra for products from 'humanely-managed' systems.

The indication is that, although the consumer's commitment to the animal welfare cause may be strengthening, it is still influenced by the retail price of the product, and this will be examined further in Section 6.4. It has to be said, however, that the unwillingness of many consumers to match their demands for high welfare standards on British farms with their own purchasing habits remains a source of irritation and anger for many farmers. Whilst recent welfare regulations have increased on-farm production costs - for example, the prohibition of sow tethers - UK producers have still had to compete with cheap imports from countries using production methods that would be illegal in Britain. As has been expressed in the NFU campaign in 1997 to 'Keep Britain Farming', UK agriculture can only survive when there is a level playing field.

Whilst this situation can only be ultimately resolved by Government intervention, I believe that a public boycott of all agricultural commodities not produced to UK welfare standards would not only express support for British farmers but also for welfare-friendly production systems (NFU Magazine, 1998). At the moment it would appear that the commitment of many people to the cause of animal welfare is not sufficient for them to take such action.

Despite this, there can be no doubt that since the 1960s there has been a growing concern for welfare issues. Several reasons may be given for this, including the increased urbanisation of the population and their subsequent distancing from contact with agricultural production, the influence of the media and the current high profile achieved by animal rights and welfare pressure groups. These will be discussed later (see Chapter 4), but, in identifying the issues, it is important to recognise the growth of a 'welfare consciousness' over this period of time in political circles as well as amongst the general public.

The UK Labour Party, for example, adopted animal welfare as an election issue in May 1997. This was highlighted in a report in 'Agscene' (Autumn 1996:12-13), of a speech delivered by Tony Blair. In this he stated that 'Giving priority to the welfare of animals is the mark of any civilised society.' Further evidence of the current status of animal issues on the political agenda was provided in the same magazine in a report, which stated that in 1996 a number of MPs from across the political spectrum had endorsed CIWF's Manifesto for Farm Animals ('Agscene', 1996:12-13).

None of this would have come as a surprise to members of the Farm Animal Welfare Council (FAWC, 1993:2), which noted in 1993 that there was an 'ever-increasing pressure on the Government and the European Commission to recommend or legislate for changes to current farming practices in order to improve animal welfare....Welfare is assuming a higher profile in the public eye and is influencing purchasing habits.'

Following the economic implications of the NFU and RSPCA surveys described above one might question the second of these observations. The pressure to legislate on welfare issues is, however, undoubtedly very real right across the European Community. Brooman and Legge (1997:179) noted the 'increased influence' of the welfare debate in EU affairs over recent years and how 'many decisions, directives and regulations have been introduced by the EU regarding farm animals'.

A further factor in the developing 'welfare consciousness' has been the number of scares involving food that have occurred during the 1980s and 1990s, from the salmonella in eggs incident in 1988 to the BSE crisis in the beef sector in 1996. Both brought livestock production into the media spotlight (Pierce, 1996) and whether the focus was on conditions in battery houses, the feeding of animal protein to herbivores, or on the environment in which animals were slaughtered, the net result was the same in that the non-farming public were confronted with these unfamiliar and more unpleasant features of the animal industry. It can only be expected that sensitivities were offended and consciences stirred as people became more aware of the welfare concerns arising out of these issues.

I believe that the farming community has largely responded in a positive fashion by making itself more open to public scrutiny and accountability. This is illustrated by such initiatives as the NFU's 'Welcome to the Countryside' programme promoting farm open days and farm assurance schemes such as the RSPCA's Freedom Foods'

(RSPCA, 1995), which promotes animal products from welfare-friendly systems. Indeed, I suggest that as a result of such initiatives UK farmers have achieved increased public support as well as a greater public awareness of and interest in farming issues.

Furthermore, I believe that such expressions of the agricultural industry's desire to renew public confidence in livestock production indicate that UK farmers not only take welfare issues seriously but that they genuinely care for their stock.

2.4   The protagonists in the debate

It is generally accepted that over recent years society has become more aware of animal welfare issues (Midgley, 1986:11), and consequently more concerned about the well-being of livestock and the systems used to breed, rear and slaughter them. Despite this, however, there is still a wide divergence of views and attitudes regarding the whole subject of their welfare. The range of views covered can be broadly categorised into three groups.

Group A: Those who use the language of rights to address issues of animal welfare. Working from the premise that animals possess rights they advocate the ending of all human activities which arguably violate those rights. For some this may simply be a matter of believing that animals have a *prima facie* right not to be harmed. So Clark (1997:11) has argued that 'animals have rights in the sense that it is wrong to multiply their suffering, that their pain is an evil and their pleasure a good.' Others, however, would consider such a definition of animal rights to be too limited and Roltson (1988:47) has identified a number of alternative interpretations. Elton has argued, for example, that animals have 'the right to exist and be left alone', Naess that they have 'the right to live and blossom', and Lilly, who spent many years trying to communicate with dolphins, declared that he 'no longer wanted to run a concentration camp for his friends' and that he believed they had 'the right to be free' (Rolston, 1988:47).

Although the origins of the concept of animal rights can be traced back to the seventeenth and eighteenth centuries, its popularisation in the second half of the twentieth century owes much to the work of Tom Regan (Leahy, 1996:189) and the publication of his book 'The Case for Animal Rights' (1984). Believing that animals, like humans, are 'subjects-of-a-life' Regan argues that they too have inherent value, i.e. 'value in themselves'. This is 'to be understood as being conceptually distinct from the intrinsic value that attaches to the experiences they have' and 'are receptacles of' (Regan, 1984:235). This 'receptacle' theme is then used to illustrate his definition of inherent value as he argues that 'it's the cup, not just what goes into it, that is valuable' (Regan, 1984:236).

Regan's contention is that 'those who have this kind of value have a valid claim, and thus a right, to treatment respectful of that value' and to have it in equal measure (Regan, 1984:267). This I believe is the most widely understood and accepted view of animal rights leading to the conclusion that 'like us.... any harm that is done to them must be consistent with the recognition of their equal inherent value and their equal inherent *prima facie* right not to be harmed' (Regan, 1984:329).

This support for the equal treatment of human and non-human 'subjects-of-a-life' leads Regan to the unequivocal conclusion that animal rights and vegetarianism are a common cause. The acceptance of the former makes the latter obligatory on the ground that the killing of an animal is the greatest deliberate, pre-meditated harm that can be done to it (Regan, 1984:351). As a result the aim of the animal rights advocate can be nothing 'less than the total dissolution of commercial animal agriculture as we know it, whether modern factory farms or otherwise' on the ground that the ownership and killing of a subject-of-a-life are unacceptable. Traditional methods of farming are seen as a compromise. Efforts to improve the welfare of animals are considered a hindrance and an irrelevance to their main aim. 'Not only are the philosophies of animal rights and welfare separated by irreconcilable differences ... the enactment of animal welfare measures actually impedes the achievement of animal rights. Welfare reforms, by their very nature, can only serve to retard the pace at which animal rights' goals are achieved' (Regan, 1984:345). The principle of animal rights is given priority over all other concerns regarding the human/animal relationship (Regan, 1984:338) and consequently the environmental, social and economic implications of bringing an end to livestock production are considered to be of secondary importance.

Group B: Those who, in contrast to the convictions of the animal rights advocates, believe that the use of animals is morally acceptable as long as the users recognise that such usage makes them responsible for the welfare of the animals concerned (Clough and Kew, 1993:xiii). Frequently motivated by sympathy and compassion as well as a sense of duty, such individuals have largely been responsible for the founding and ongoing work of many of the animal welfare and protection organisations, including the Royal Society for the Prevention of Cruelty to Animals (RSPCA).

Rather than follow Regan's belief in animals and humans sharing an equal inherent value (Regan, 1984:329) welfarists are more likely to accept and live by what Rolston defines as the 'traditional ethic' concerning the human/animal relationship (Rolston, 1988:45): 'Use animals for your needs, but do not cause needless suffering.' Such an ethic is 'prohibitive on the one side, enjoining care of domestic animals to prevent suffering. It is permissive on the other side; subject to prohibited cruelty, animal good may be sacrificed for human interests' (Rolston, 1988:45). However, as an ethic it is clearly open to many different interpretations, especially with regard to how 'needless suffering' is to be defined. As a result, a wide range of different views regarding the treatment of animals is found amongst welfarists, particularly with regard to livestock production systems.

For example, some welfarists will reject all intensive farming of animals in the belief that such methods impose unnecessary suffering on the animals concerned (Harrison, 1964). As a result they will only purchase animal products which they believe to have been reared in humane production systems, from either organic and/or extensive farms. Where there is any uncertainty about the source of animal products then they may well choose to reject those products in favour of a vegetarian lifestyle. Others, however, whilst acknowledging that some features of intensive systems may impose needless suffering on the animals, are ready to argue that once those features are improved there is no reason to disallow those systems on welfare grounds. Whilst, for example, many welfarists argue that the battery production of eggs should be banned, others contend that an acceptable compromise can be achieved with those wishing to employ such systems when improvements are made to the size, design and stocking density of battery cages (Webster, 1994:261).

Most of the mainstream UK farming organisations would support the second of these views. This is indicated in the NFU's statement on farm animal welfare in which it is acknowledged that 'farm animals are sentient beings; for example they can feel pain, fear and fatigue. But we reject the attribution to farm animals of human characteristics and perceptions - anthropomorphism - and of the same rights and responsibilities as human beings. Nevertheless humans have a clear duty to treat all animals humanely and to refrain from causing cruelty and unnecessary pain. The NFU is committed to the humane treatment of farm animals ... and endorses the principle set out in the NFU's 1991 policy statement that decisions on animal welfare must be consistent with economic and practical reality' (NFU, 1995:3).

Group C: The very small minority of animal users and livestock producers who believe that they have the right to use and abuse animals for their own ends, regardless of the effect that those ends have on the animals, especially when those ends are related to increased profit margins. As an illustration of this sort of attitude Midgley (1983:10) quotes a television interview in which a scientist declared that 'he did not think that experimenting on animals raised a moral issue at all.' Midgley (1983:10) suggests that people who express this view are likely to believe that animals 'fall outside the province of morality altogether ... and that claims on behalf of animals are not just excessive, but downright nonsensical, as meaningless as claims on behalf of stones or machines or plastic dolls'.

2.5   The problems of language

Because the subject of animal welfare is likely to evoke a great deal of passion from representatives of all the above groups it is inevitable that, as they engage in the debate on the related issues, they are likely to resort to using emotive language. This was apparent for example in the quote in Section 2.2 from Rachel Carson's Preface to 'Animal Machines' (Harrison, 1964:vii). As will be argued at a later stage, such language can easily result in a loss of both objectivity and clarity in an argument, and may obscure and obstruct rational thought.

It should be appreciated, however, that emotion does not have a monopoly in irrational and imprecise argument, especially when it comes to the debate on animal welfare. Much of the language used in the debate is less than precise. Concepts such as suffering, cruelty, pain, and stress are not only awkward to define, but when applied to animal species the experiences they describe are in themselves difficult to measure, as 'there are few positive methods of assessing the well-being and contentment of animals' (FAWC undated: 2-3).

Similarly, whilst all reasonable people are likely to believe that physical cruelty is offensive and morally indefensible, it can often be the case that practices deemed cruel by one person may be considered perfectly acceptable by someone else. When such concerns are then applied to animals, and especially to a variety of husbandry systems, the issues become even more complex and one is faced with questions, for example, as to whether boredom is cruel, or whether the imposition of an unnatural environment is morally exploitative. As animals are unable to express their opinions on any matter in a direct sense, the confusion and lack of precision can only be magnified. These and related issues will be examined in detail in Chapter 5.

Definitions

Several definitions are required in order to clarify the foregoing and following discussions. For example, reference has already been made to both 'intensive' and 'extensive' farming and also to 'animal welfare'. As these concepts are integral to the discussion it is important that they now be defined as clearly as possible.

Animal Welfare
Whilst animal welfarists tend to be concerned with the well-being of all species, rather than focusing purely on farm animals, there are some definitions that interpret welfare in terms that are specific to agricultural livestock. Sainsbury (1986:3) provides one when he

describes good welfare in terms of 'a husbandry system appropriate to the health and, so far as is practicable, the behavioural needs of the animals and a high standard of stockmanship.'

The Report from the Brambell Committee (HMSO, 1965:9) suggested that animal welfare 'embraces both the physical and mental well being of the animal. Any attempt to evaluate welfare therefore must take into account the scientific evidence available concerning the feelings of animals that can be derived from their structure and functions and also their behaviours.'

This appreciation of both the mental and physical needs of the animals is now widely accepted as the basis for any consideration of the welfare needs of any animal. Indeed FAWC (1993:3) suggests that 'the welfare of an animal includes its physical and mental state and we consider that good animal welfare implies both fitness and a sense of well being'. What this means is then defined in terms of the 'Five Freedoms':

Freedom from thirst, hunger and malnutrition
Freedom from discomfort
Freedom from pain, injury or disease
Freedom to express normal behaviour
Freedom from fear and distress

FAWC describes these five criteria as 'ideal states rather than standards for acceptable welfare. Nevertheless they form a logical and comprehensive framework for analysis of welfare' (FAWC, 1993:3). The implications for the UK livestock producer, as a result of these standards being increasingly accepted as the reference point for farm animal welfare by industry and consumer alike, will be discussed later. Suffice it to say at this point, however, that the widespread adoption of the 'Five Freedoms' would inevitably lead to the demise of a number of intensive production practices and systems such as the use of battery cages in the poultry industry.

Intensive Livestock Farming
It is easy to describe intensive farming in a very emotive and imprecise way, as has been illustrated in Section 2.1 with Compassion

in World Farming's comments on such systems. A clearer, more objective and therefore, for these purposes, more acceptable definition, is provided by the Brambell Report in which intensive husbandry is described in terms of those production methods which 'result in the rapid production of animal products by standardised methods involving economy of land and labour...a high degree of mechanisation and automation is a feature of many of the larger establishments' (HMSO, 1965:2). Such systems may well raise ethical questions with regard to how animals are perceived as well as treated but, as has already been indicated, the definition of intensive farming does not assume that such systems are necessarily cruel.

Extensive Livestock Farming
Extensive farming is commonly perceived as being the alternative to intensive production, whereas examples of both management systems frequently co-exist on the same farm. A familiar pattern in livestock rearing areas is the intensive dairy unit, which incorporates an extensively farmed beef enterprise.

A definition of extensive agriculture, which I believe would receive wide assent within the farming community, is provided by the Universities Federation for Animal Welfare (UFAW 1993). Extensive systems are those which 'tend to release the animals from the closely confined, highly controlled environments found in intensive systems, but still make use of prophylactic veterinary medicines; in effect extensive livestock systems may include animals kept under all sorts of conditions; free range, in and/or out of housing, and products may be marketed with a wide variety of health or animal welfare labels.'

Ethics and Morals
Although the two words have different origins ('ethics' is derived from the Greek *ethikos* and 'morality' from the Latin *moralis*) they 'have come to be treated as almost identical in meaning' (Vardy and Grosch, 1994:14) and, as in this study, are used interchangeably.

As regards the definition of morals, Jenkins (1969:165) states that the word 'is taken as meaning the actual conduct commended and practised as a matter of fact. Morals has to do with the decisions taken and the advice given in matters of conduct, individual and corporate.' Put simply, morals are 'concerned with right and wrong conduct' (Collins Dictionary 1996).

2.6   The need for an ethical basis for livestock production

As in every other walk of life, the farmer is constantly faced with situations that have both ethical and economic implications. In some cases he may be able to satisfy his conscience and his accountant, in others there will be a conflict that he will only resolve by favouring one at the expense of the other. In this his situation is no different from that of any other businessman.

The resolution of the farmer's conflict is rarely so simple, however, in that a number of other factors have a bearing on the outcome. Farming, for example, is probably more subject to the vagaries of political interference and manipulation than any other industry. This may well result in the producer being legally forced to engage in practices that offend his sense of responsibility towards his stock and which consequently conflict with his ethical beliefs and principles. This was illustrated in 1999 when the complexity and expense of further meat inspection legislation brought about the closure of many small abattoirs. As a result producers had to transport their animals greater distances to be slaughtered, thus causing the animals further discomfort and stress. The subsequent anger within the farming community has been expressed in a number of ways, not least in widespread representation to the Ministry of Agriculture Fisheries and Food (MAFF) demanding that the legislation be changed and local abattoirs reconstituted.

A further factor is that the livestock producer is very much at the mercy of the retailers and consumers. Again, the producer's situation is little different from that of any other business, only in this case the moral issues are to do with living sentient beings about which the general public have strong, if sometimes misinformed, opinions. As UK pig farmers have discovered to their cost, whilst the consumer might on the one hand demand high standards of animal welfare from British producers and support legislation that brings this about, they are still likely to purchase imported animal products reared in cheaper and crueller systems which are outlawed in their home country.

On the other hand, whilst a farmer of his own volition may want to employ production methods he believes to be better for his stock, he may well be prevented from doing so by consumer resistance. This may be due to consumers being unwilling to pay the extra cost, or it may stem from their lack of agricultural knowledge and their opposition to systems and practices that they do not understand. I believe that this second situation arose recently when the RSPCA expressed their opposition to the use of farrowing crates in pig units. Many producers would argue for the retention of stalls on welfare grounds, believing that they protect the life of the young piglet whilst causing minimal stress to the sow. Now, as a consequence of the RSPCA's stance on the use of farrowing pens, any pig farmer wanting to join the organisation's 'Freedom Foods' assurance scheme would find that the scheme's criteria either prevented him from joining or forced him to dispense with farrowing crates to the possible detriment of piglet welfare. This subject is discussed in more detail in Chapter 6.

In all of these cases the producer is faced with a variety of complex ethical issues that affect the economic viability of his farm. At the same time he has to contend with a multiplicity of factors that influence how he addresses the issues and makes an economic or ethical judgement. In addition to all of this he is confronted by a variety of vociferous individuals and organisations that not only oppose what he does for a living, but who believe that his use of animals is cruel and morally indefensible. Indeed, their expressed aim is to put him out of business (Regan, 1984:351).

In direct contrast to Regan's view I believe that the vast majority of UK livestock producers care deeply about their animals and that the business in which they are engaged is not only morally defensible but is honourable and worthy of support. Whilst such support is often asked for and given in terms of purchasing British agricultural products, my contention is that it is also required in terms of meeting the arguments of the critics and opponents. For this to happen there has to be an argued case for the ethical principles that undergird the UK livestock industry, and especially its use of intensive systems.

What is required is the establishment of an adequate ethical principle with appropriate criteria that will enable a system, method or practice to be assessed ethically.

In the current agricultural crisis, one key to the industry's survival must be the ability of its practitioners to engage in the art of apologetics, demonstrating 'the reasonableness of (their) position in the light of modern scientific and philosophical views' (Lord and Whittle, 1969:10). The purpose of this study is to help provide this key by establishing an ethical principle for intensive livestock production systems along with the criteria by which the ethical acceptability of those systems can be assessed.

In the chapters that follow, the historical development of intensive livestock production is described with particular emphasis on the relationship between man and animal (Chapter 3). The current non-farming view of intensive livestock production is then evaluated critically with an analysis of the implication for the future of these husbandry systems (Chapter 4). The philosophical and theological issues involved in animal/human interactions are also considered in Chapter 4 and these are further discussed in Chapter 5 with regard to the establishment of an ethical principle for assessing the acceptability or otherwise of contemporary livestock production methods. Finally, the practical implications of the adoption of the ethical principle are examined in Chapter 6, followed by discussion of the economic implications in Chapter 7.

CHAPTER 3

The Historical Development of Livestock Ethical Issues

In his paper 'What is good agriculture?' Freudenberger (1994:47)
makes the comment that 'historical analysis will be an essential
element of agricultural ethics issues in the future.' In the context of
his paper, the author is referring specifically to arable farming and the
ecological challenges being faced by that sector of the industry,
although his words are equally applicable to the ethical concerns
arising out of certain practices and systems currently being used in
UK livestock production. The implications of such historical analysis
are emphasised later when he states that as a consequence 'we will be
better equipped to define more clearly the agenda in agricultural
research and development for the twenty first century' (Freudenberger,
1994:47). He is clearly suggesting that it is only when we understand
our agricultural past are we able to contextualise the present and thus
take account of those influences that have moulded current attitudes
and practices. Only then can sound judgements be made about the
credibility of those attitudes and practices and about the subsequent
decisions that have been made concerning their future application and
use.

This chapter examines the historical relationship between animals and
humans as it appertains to livestock husbandry, and considers the
implications of that history for the development, use and acceptance
or criticism of intensive livestock production systems. It should be
noted that the examination will be restricted to those influences that
have affected the development of European rather than global
agriculture, as it was largely from them that current farming practices
in the United Kingdom evolved.

3.1 Early references to the human/animal relationship

From their examination of a number of primitive humanoid tooth and jaw fragments taken from various archaeological sites, Leakey and Lewin (1992) have been able to hypothesise the evolutionary development of the human diet. Whilst the earliest fragments examined indicate that our most distant ancestors were vegetarians other evidence indicates that they soon acquired a taste for animal flesh. 'With the origin of the *Homo* lineage, the trend toward bigger grinding molars became reversed, not to fruit processing teeth again, but to the teeth of omnivorous animals that may have included meat in their diet' (Leakey and Lewin, 1992:54). Although these views have not been free of criticism and opposition, the authors consider their claim to be well founded and Leakey has unequivocally stated that 'although some anthropologists argue that regular meat eating was a late development in human history, I believe that they are wrong. I see evidence for the expansion of the basic omnivorous hominid diet in the fossil record, in the archaeological record and, incidentally, in theoretical biology.' (Leakey and Lewin, 1992:165).

If Leakey's views are correct then it would seem that there is sufficient evidence to believe that the desire or need to eat meat has been a fundamental element in the relationship between humans and animals almost since the dawn of our history as a species. How this was manifested in terms of the relationship, however, remains the cause of some contention. Whilst recognising that the accepted view of the hunter/gatherer is of a food collector living in an antagonistic relationship with the rest of the animal kingdom, Ingold (1994), for example, suggests that the evidence available presents a very different picture. Rather than antagonism, there appears to have been a deep sense of respect that verges on awe if not actual worship.

From observing the behaviour and attitude of contemporary hunter/gatherer societies, Ingold (1994:9) postulates that such groups treat 'the country, and the animals and plants that live in it, with due consideration and respect, doing all one can to minimise damage and disturbance....One treats an animal badly by failing to observe the proper respectful procedures in the processes of butchering, consumption and disposal of the bones or by causing undue pain and suffering to the animal in killing it.' The hunter knows his prey, identifies with it and empathises with it to the point of experiencing the sense of respect and awe exhibited in early cave paintings.

This is the picture of the hunter/hunted relationship that Serpell and Paul (1994:131) also described when they wrote of the way in which 'hunting mythology depicts an egalitarian community of animals and people bound together by an unwritten contract of mutual respect and dependence; a sacred pact between predator and prey in which the interests of neither party are viewed as subordinate.' This perception of the hunter's outlook on his world is then compared and contrasted with that of the farmer and the pastoralists who, it is suggested, out of necessity, 'set themselves up in opposition to nature' (Serpell and Paul, 1994:131). The assumption being made is that the transition from a wandering hunter-based existence to a settled life based on farming marked a fundamental change in man's attitude, both to the world in which he lived and to the creatures with whom he shared it.

This crucial evolutionary step in the human/animal relationship took place approximately 9000 years ago. At some point our distant forebears discovered that it was easier to domesticate animals and breed them for meat, rather than pursue them as game in the frequently futile hope of obtaining a kill and thence a meal. Beginning with goats and sheep our early attempts to breed and rear livestock were so successful that cattle, pigs and horses were soon added to the domesticated list (Clutton-Brock, 1994:27).

Like Serpell and Paul, Ingold (1994:16) believed that domestication represented a fundamental change in the human attitude towards other species and accordingly wrote of how 'the relationship of pastoral care, quite unlike that of the hunter towards animals, is founded on a principle not of trust but of domination, of masterly control ...The transition in human-animal relations that in western literature is described as domestication of creatures that were once wild, should rather be described as a transition from trust to domination.'

Ingold's conclusions are clearly of interest to those who view modern livestock systems with disdain and distaste and who believe that current husbandry methods deny the essential dignity and nature of the animals being farmed. As will be discussed later, many who hold such views also tend to believe that in domesticating livestock mankind has dispossessed the said animals of some of their most basic rights. They have been treated as commodities that exist purely for our use and pleasure with little regard paid to their sentiency and even less to our responsibility. The validity of this line of argument will be discussed more fully in Chapter 4, but I would suggest that it might well have influenced much of Ingold's own thinking.

The reason for making this suggestion is that I believe it is equally possible from the evidence available to come to very different conclusions from those outlined above. For example, rather than being an example of man's exploitation and domination, as Ingold has argued, the domestication of livestock may well have been part of a natural evolutionary process, which was of mutual benefit to both parties. As one provided for the other in terms of food, clothing, and companionship, so the arrangement was reciprocated in terms of protection and security of food provision. Rather than being a relationship of domination and even antagonism, it could equally have been one of mutual trust and benefit.

This is the view expressed by Budiansky (1994:24), who stated that 'domesticated animals chose us as much as we chose them' and that behavioural and archaeological evidence 'points to the co-operative evolution of our species in a mutual strategy for survival.' He then suggests: 'that leads to the broader view of nature which sees humans not as the arrogant despoilers and enslavers of the natural world, but as part of that natural world and the custodians of a remarkable evolutionary compact among the species.' This 'evolutionary compact' is defined as 'a finely honed evolutionary strategy for survival ... In a world made up of so much competition for survival, nature has with surprising frequency cast upon the solution of co-operation.'

This interpretation of the development of the man/animal relationship in full accord with the concept of man's dominion over animals as it is detailed in Genesis 1:26. This describes how 'God said "Let us make man in our image, after our likeness; and let them have dominion over the fish of the sea, and over the birds of the air, and over the cattle, and over all the earth, and over every creeping thing that creeps upon the earth."' The concept of 'dominion' over the rest of creation is here compared with the relationship between that creation and its Creator. An acknowledgement of man's responsibility under God to exercise a similar rule over all other forms of animal life is implied - a rule based on respect and care, and not on exploitation, antagonism and abuse. In return, the animals provide man with the answer to many of his physical needs in terms of food and clothing.

I believe that this is what is being suggested by Von Rad (1956:58) when he states that 'man's creation has a retroactive significance for all non-human creatures; it gives them a new relation to God. The creature, in addition to having been created by God, receives through man a responsibility to God; in any case, because of man's dominion it receives once again the dignity belonging to a special domain of God's sovereignty.' This understanding of the biblical concept appears to be further supported by a number of other references in the Bible concerning the human/animal relationship, e.g. Deuteronomy 22:6 and 25:4. In these and similar passages explicit directions are given as to how man's dominion over other species must be exercised through the acceptance of a God-given responsibility of care towards them. The writers believed that man is not the master of creation but its steward, with a direct accountability to the Creator for its care and well-being. In referring to Psalm 8, Jones (1988:18) clearly comes to a similar conclusion when he suggests that 'mankind is seen as a kind of managing bailiff for God. An agent with authority to use available resources to support all those dependent on the estate, but bound to give account of his use and custody of what is not his own.'

Some of the practical implications arising out of this sense of responsibility for the care of God's creation are clearly demonstrated in the words of Psalm 23. This piece of Hebraic hymnody vividly and familiarly portrays the responsibilities of the shepherd who cares for the animals in his charge by feeding and watering them (v.2) by protecting and healing them (v.4 and 5) and by putting their safety before his own (v.5). Indeed, such is the degree of commitment expected of the shepherd that Christ later compares his own impending crucifixion with the sacrifice of the good shepherd who 'lays down his life for the sheep' (John 10:15).

The influence of this biblical concept of responsible stewardship, especially in terms of the man/animal relationship, cannot be underestimated - it is at the root of many of the traditionally accepted standards of animal care and stockmanship. Furthermore, I also suggest that this has largely come about as a direct result of the major role which the Church has played in the development of English agriculture. This role, which was particularly evident in the improvements in livestock breeding and care initiated by the medieval monastic orders, will be discussed more fully in the next section.

From the above I would argue that the awe and respect of the hunter for his prey, to which Ingold draws attention, was not lost with the domestication of animals. Rather, there is evidence to suggest that as tribes and nations settled to a more agriculturally-based existence the opposite actually happened. This can again be illustrated from the Old Testament text in which there are a number of references to cattle, and especially bulls, being given such a major religious significance that they became the focus of worship and the symbol of divine attributes (e.g. Exodus 32, Hosea 8). Indeed Schwabe (1994:36) goes as far as to suggest that the earliest domestication of cattle was a result of cultic development rather than food provision. He also suggests that it was their religious significance that caused them to be valued as an early source of primary wealth (Schwabe 1994:41).

This place of animals within the cultic life of a nation had a marked effect on a number of civilisations, including that of Rome where the religion of Mithras the bull was particularly significant. As the empire expanded so one can assume that the very nature of colonial life would have encouraged the subsequent spread of this religious influence to subject races and nations, including England, and the subsequent enhancing of the place of cattle within those societies. Indeed, Marcus Terrentius Varro, who produced a significant work on Roman imperial agriculture, (quoted by Schwabe, 1994:46) indicates the cultic value placed on cattle when in describing how copper coins were marked with their insignia, commented that 'the very word for money is derived from them, for cattle are the basis of all wealth.' I contend that factors such as these indicate that Budiansky is correct and that the value of cattle was enhanced rather than diminished by domestication. Rather than being debased, as some would suggest, there is every indication that domestication led to certain species being increasingly revered.

It is important to recognise, however, that man's appreciation of the value of animals and his dependency on them was not confined to the areas of food, clothing, and religion. Seebohm (1976:37) for example, reminds us that 'the man who discovered that the fierce bull could be converted by emasculation into the patient and tractable ox was the bringer of a great gift to mankind.' That one discovery led to stockmanship being a key feature in arable as well as livestock farming, as first oxen and then horses were used for traction, and this is how it was to remain until well into the twentieth century. Seebohm (1976:46) also describes how the Celts, like the Romans, used cattle as their economic base. 'Cows among a primarily pastoral people such as the Celts were of the greatest importance, not only on account of their milk, flesh, hides and power of traction, but as the recognised standard of wealth.'

It is clear that many agricultural societies throughout history have used their livestock for a wide range of purposes and have valued them for an equally wide variety of reasons. It would be surprising, therefore, if as Ingold (1994:16) suggests, the relationship between farmer and stock were one of antagonism and domination. Rather, one would expect that those who farmed animals would recognise this state of mutual dependency, and consequently respect, and behave responsibly towards, the stock in their care. Furthermore, if this were the case then one would expect to find such an expression of responsibility and compassion towards all domesticated animals reflected in the earliest agricultural literature, as indeed it is. Seebohm (1976:63) cites the Roman writer Giraldus Cambrensis, for example, who advised ox drivers to drive their animals gently 'so as not to break their hearts.' Likewise, he quotes from Palladius, another Roman writer whose works were translated into English in the fifteenth century who, in describing what he considers to be adequate stock housing, states that 'stables should look south with a north light and should have a fire in winter' to warm the beasts (Seebohm,

1976:77). He also suggests that oxen should be shoed with 'old broom' both to protect their feet and to act as a charm against weasels! Columelia, a writer from the first century AD offers the following very sensitive advice to shepherds: 'The deliverance of a pregnant ewe should be watched over with as much care as midwives exercise, for this animal produces its offspring just in the same way as a woman, and its labour is painful even more often since it is divorced of all reasoning' (cited by Clutton-Brock, 1994:34).

Such writings set the tone for much of the guidance that is given to livestock producers in subsequent centuries. The emphasis is on responsible care, which in turn is based on due regard for the dignity and well-being of an animal that is valued and respected. Moreover, it is these principles that have been traditionally defined in the concept of stockmanship. The early English document 'Colloquium of Aelfric' includes a description of the duties of a stockman, and emphasises his role in protecting his animals from all that may threaten them (Seebohm, 1976:125). In the thirteenth century, there are various references to the work of a shepherd in recognising and treating diseases amongst his flock, including sheep scab and foot rot. The suggested medications for scab include an ointment made from quicksilver and lard, or a tar dressing with butter or lard (Seebohm, 1976:130). In line with much of Seabrook's work on the personality profile of the stockperson which is examined in Chapter 5, the thirteenth century writer Bartholomew Angelicus states that an ox driver 'must not be melancholy or bad-tempered, but gladsome, singing and joyful', whilst Walter of Henley in the same vein states that a waggoner was 'expected to be steady, experienced, modest and not wrathful, skilfully directing his horses and not over-burdening them' (quoted by Seebohm, 1976:143).

Later agricultural commentators continued along the same line. The sixteenth century writer Thomas Tusser, quoted by Evans (1956:43), set his advice in verse, and included the following directions to shepherds in his 'June Husbandrie':

> 'Wash sheep for the better – where water doth run,
> And let him go cleanly and dry in the sun;
> Then shear him and spare not at two days an end,

The sooner the better his corps will amend.
Reward not thy sheep, when ye take off his coat,
With twitches and patches as broad as a groat;
Let no such ungentleness happen to thine,
Lest fly with the gentils do make it so pine.'

It has to be said that Tusser's own attempts at farming were a dismal failure and he ended his days in a debtor's prison. Other contemporaries, such as Fitzherbert and Barnaby Googe, however, not only provided sound counsel, but were also able to prove the worth of that counsel through their own example as successful farmers. Googe in particular had a major role in the development of English agriculture in that he is credited with the introduction of clover and turnips as stock feed in this country. Having spent some years working with and learning from farmers in Holland, he then applied those lessons on his return home and described them in his 'Foure Books of Husbandrie' published in 1577. The application of ideas and methods such as these brought about major improvements for the health and well being of English livestock, despite the fact that initial opposition meant that it was some time before they received widespread acceptance.

As the author of the 'Boke of Husbandry', Fitzherbert was the first author since Walter of Henley 200 years before to produce a major work on English agriculture (Seebohm, 1976:206). Basing much of his advice on the wisdom of previous generations his aim was to promote profitable and efficient farming practice. This can be illustrated with reference to his comment that 'it is an old saying: he that has both sheep, swine and bees, sleep he, wake he, he may thrive. And that saying is because they be those things that most profit rises (from) in shortest space with the least cost' (Fitzherbert, 1523, quoted by Addy, 1972:65). Furthermore, by emphasising the importance of practical husbandry Fitzherbert was instrumental in bringing to an end many of the old superstitions regarding matters such as livestock breeding. The old idea, for example that the sex of an animal's progeny could be determined by arranging for the mating to take place during the waxing or waning of the moon, was dismissed as nonsense and replaced with sound advice on practical breeding management.

In many ways Fitzherbert epitomised the spirit that pervaded English agriculture during the sixteenth and seventeenth centuries, which made it a major period of agricultural expansion and improvement. This was effected partially through the works of writers such as Fitzherbert and Googe, but it was also influenced by the growing support for and interest in land enclosure. Whilst enclosures were to have many disastrous social ramifications, they were also to prove one of the most influential factors in the improvement of livestock breeds and in the overall condition of farmed animals in England. As long as the ancient open-field system that had typified English farming for centuries continued in use, livestock husbandry suffered, as animals were subjected to the constant threats of widespread infection and indiscriminate breeding.

The Enclosure Movement changed all this, and the impact it made on English agriculture can be illustrated by the estimated figures for the period 1760 to 1820, when approximately four thousand separate Enclosure Acts were passed by Parliament. These resulted in over five million acres of common grazing and open fields being enclosed (Richards and Hunt, 1965:5). The procedure for bringing about the enclosure of an area involved a group of local people promoting a Bill through Parliament. This normally required the agreement of the lord of the manor, the tithe owners and the owners of four-fifths of the open field system. As the wealthy obviously owned most of the land, the decision to promote an Enclosure Bill would often be taken by a few individuals, with the majority of the poorer small farmers being excluded from the process. Consequently there were many abuses of the system as well as a great deal of impotent anger from those who suffered as a result of it, especially when the General Enclosures Act of 1801 rendered individual Bills unnecessary. As Richards and Hunt (1965:7) recognised 'The necessity for enclosure was undeniable, and its purely agricultural results were excellent; but the social consequences of the way in which it was carried out were often deplorable.'

As has already been illustrated, the principles of stockmanship had been expounded for many centuries but it was only with the advent of land enclosure that they could be fully implemented. Livestock welfare was no longer subject to an agricultural system that allowed diseased and healthy animals to mix together and inappropriate breeding to take place. At the same time, as has already been recognised, there was the general casting aside of old superstitions in favour of a more reasoned approach to life. The age of science was dawning and with it a greater understanding of the needs of animals as well as of their productive capabilities. This tended to re-emphasise rather than demote traditional principles of care as is indicated in some words from the sixteenth century, offering advice to shepherds: 'The shepeard (must) have a little board sette fast to the side of his little fold to lay his sheepe upon when he handleth them and a hole bored in the board with an auger and therein a grained stake of two feet long to be set fast to hang his tarre box upon so that it shall not fall. And a shepeard should not go without his dogge, his sheep hooke, a pair of sheares and his tar box' (quoted by Evans, 1956:40) The emphasis of the anonymous writer is on a professional efficiency and expertise that is directed to the care and welfare of the flock, with all the tools of the trade being readily available for that sole purpose. His words exemplify a balance between stockmanship and husbandry, a combination of skills and attitudes, which is endemic to those systems of livestock farming that are concerned for both animal welfare and profitability.

Through the centuries, the indications are that the traditions of stockmanship and husbandry that were being steadily established were founded on principles of responsible husbandry expressed in terms of respect and regard for the animal rather than domination and exploitation. These principles were to provide the foundation for the continued improvements in the health, welfare and productivity in the flocks and herds of eighteenth and nineteenth century England, through the continuing Agricultural Revolution and through the subsequent scientific and technological advances that were yet to come.

3.2 Theological and philosophical developments

Mention has already been made of how biblical teaching influenced the development of the human/animal relationship in western society through the teaching and example of the Church. Such teaching has largely been based on the concept of human 'dominion' over the rest of creation as described in Genesis 1:26. Rather than encouraging man's exploitation of the rest of the animal kingdom, as some suggest, I would argue that this concept actually sets demanding standards of human responsibility towards other species. Based firmly on an understanding of dominion as exercised by God over his creation it has more to do with responsibility and duty rather than licence and exploitation.

Something of this is illustrated in the number of biblical commandments, which emphasise the human responsibility towards domesticated stock. Exodus 20:10, for example, directs that cattle as well as their owners should be allowed a sabbath day's rest. In a similar vein, whereas people are subject to rigorous sabbath day restrictions, animals are exempt. As a consequence they should be rescued if they are trapped, treated if they are hurt (Luke 14:5) and have water and food taken to them as when they are hungry or thirsty (Luke 13:15). Other laws relating to the welfare of stock include one that directs how oxen are not to be muzzled when threshing corn (Deut.25:4), thus allowing them to eat when they are hungry, and another which declares that a hen bird with eggs or young is not to be taken (Deut. 22:6).

An even more telling reference is found in Proverbs 12:10, where kindness to animals is equated with the quality of righteousness, the very characteristic of God himself. In stating that 'a righteous man cares for the needs of his animals' the writer is suggesting that the individual who behaves in a caring way towards his stock is reflecting an attribute of the Divine. I believe that this one verse expresses the whole tenor of biblical teaching with regard to the human/animal relationship, as that relationship should be based on responsible care and use allied to sympathy and kindness.

As a result of the widespread teaching, influence and example of the Church regarding these matters I suggest that this understanding of dominion in terms of a responsibility to care for stock would have had an important influence on the development and overall concept of the principles of stockmanship. Indeed, this influence would have been particularly pronounced in this country, as it was the Church which led the way in improving livestock breeding and husbandry in medieval England. The monastic houses were not only amongst the largest landowners of the time but were also some of the most progressive farmers. As Briggs (1983:77) commented, 'monasteries sometimes led the way in good management, and it is interesting that whilst Walter (of Henley) was writing in Canterbury: Henry of Eastry, the Prior of Christ Church Canterbury was one of the most astute stock-keepers and managers of his day.' They encouraged and initiated many improvements in livestock breeding, especially with cattle and sheep, and helped to make agriculture the foundation of the nation's wealth.

In this, as in all areas of their communal life, they were guided by their study of Scripture and by the teaching of the saints, both of which they imparted to others. Unfortunately, however, the views of the some of the saints were often contradictory, and as a consequence the Church's attitude with regard to the place and status of animals was frequently ambivalent. Whilst some emphasised the Scriptural injunction to exercise responsible care towards other species, others just as emphatically dismissed it.

The history of the early church, however, is littered with stories of care being shown towards animals by many of its most saintly figures. 'St. Jerome has regularly been depicted as living, together with his tame lamb or donkey, with a lion which appointed itself to be his bodyguard after he had befriended it by getting a thorn out of its paw. St. Giles, patron saint of cripples, received his injury through defending a tame hind. The cell of the anchorite St. Theonas was constantly frequented by buffaloes, goats and wild asses (Hume, 1957:26). Hume recognises that the historicity of many of these stories may be questioned, but he also makes the point that each story emphasises that this 'was what was expected of a saint in those days.'

The father of the Franciscan tradition and the epitome of this quality of care was St Francis of Assisi (1181-1226). In the oft-quoted description provided by St. Bonaventure there are echoes of the earlier teaching of Chrysostom. 'He was filled with a greater gentleness when he thought of the first and common origins of all beings, and he called all creatures, no matter how small they were, by the name of

45

brother or sister, because he knew that they all had in common with him the same beginning' (Linzey, 1994:66). Linzey (1994:67) argues that this was in total contrast to the spirit of the times, and that Francis and those like him symbolise a radical attitude towards animals that is timeless in its relevance. 'What so many of the saints force us to wrestle with is the idea that we must view creation from God's own perspective and not our own. The worth of every creature does not lie in whether it is beautiful (to us) or whether it serves or sustains our life and happiness. Only if we can save ourselves from an exaggerated anthropocentricity can we begin to construct an adequate theology of animals'.

Linzey's views, along with those of other current writers, will be considered more fully in Chapter 4, but one is left wondering with him as to how far this spirit of Franciscan care and respect for animals permeated the life of medieval monasteries. If it was widely accepted and adhered to then it must have had a marked impact on their work with regard to the improvement of both husbandry and stockmanship. Francis clearly wanted this to be so and his desire was that his teaching should be relevant at a practical level. Within his own order, for example, his rule of life was both caring towards animals and yet pragmatic in its appreciation of human dietary requirements. He could extol the beauty of all living things and encourage a caring and responsible attitude towards them, whilst believing that when necessary they could still be used for the satisfying of human need. His was not a vegetarian creed, as is illustrated in his 'Rule' of 1221 in which 'it is carefully stated, "In obedience to the Gospel, they (the friars) may eat of any food put before them"' (Sorrell, 1988:75).

Like Linzey, Sorrell believes that the Franciscan view of life is still of relevance. 'Francis looked beyond others of his time to envision a world in which humankind shared a concern for the whole community of creatures, and he expressed his love and concern in such original and moving ways that they may still give us inspiration today' (Sorrell, 1988:144). Hume (1957:27) likewise believes that Franciscan teaching has always been of influence in the life and teaching of the Church, and that it still commands respect today.

It was not the teaching of St. Francis that was to dominate the life of the Church, however, but that of St. Thomas Aquinas (1225-74) whose theology is contained in his great work 'Summa Theologica'. It is from this that Linzey (1994:13) identifies three key features which he suggests encapsulate Aquinas's thoughts concerning the status of animals. 'First, animals are irrational, possessing no mind or reason. Second, they exist to serve human ends by virtue of their nature and by divine providence. Third, they therefore have no moral status in themselves save in so far as some human interest is involved, for example, as human property'.

These assertions stand in stark contrast to most aspects of Fanciscan teaching regarding the human/animal relationship. Linzey (1994:67) contrasts the two schools of thought by suggesting that whilst Aquinas's emphasis is anthropocentric, focusing as it does on the pre-eminence of human interests over those of other creatures, the teaching of Francis is totally God-centred with its declaration that 'only God, and not man, is the measure of all things'. It was the theology of Aquinas, however, that was accepted as the measure of faith and in 1567 he was declared 'Doctor of the Church', with the study of his works eventually becoming obligatory for all students of theology. Consequently, his views on the relationship between humans and animals have remained the accepted teaching of the Roman Catholic Church on this matter for the last seven hundred years. The first authoritative statement even to suggest that any other view could possibly be considered by the Church came in a papal encyclical in 1988, 'Solicitudo Rei Socialis'. In this Pope John Paul II urged the Church to have 'respect for the beings which constitute the natural world' and added, 'The dominion granted to man by the Creator is not an absolute power, nor can one speak of a freedom to "use and misuse," or to dispose of things as one pleases ... When it comes to the natural world, we are subject not only to biological laws, but also to moral ones, which cannot be violated with impunity' (Singer, 1995:196).

The teachings of Aquinas have undoubtedly been a major factor in influencing people's attitude towards other species, especially in those countries that have a predominantly Roman Catholic culture. Whilst he does advocate kindness towards animals in 'Summa Theologica', it is only as a means of encouraging kindness between people and not as a benefit for the animals themselves. He writes: 'God's purpose in recommending kind treatment of the brute creation is to dispose men to pity and tenderness towards one another' (cited by Hume, 1957:29). The well being of the animal is of little or no concern.

Against this background, there can be no doubt that the three elements that Linzey identifies in Thomist thinking have tended to encourage a *laissez faire* attitude towards other species which has overshadowed his instruction to show them kindness. This contrasts with Franciscan teaching in which the highest standards of welfare and care have been clearly advocated (Sorrell, 1988:142). Whilst one school encourages benevolence towards livestock, the other appears to encourage the belief that animals exist solely for human benefit and profit. Exponents of both these views are still found amongst today's agricultural practitioners and both are pivotal in the ongoing ethical debate regarding the use of intensive livestock production systems.

One who was deeply influenced by the works of Aquinas and whose philosophy was to have a profound and disturbing effect on the relationship between humans and animals was the French philosopher Rene Descartes (1596-1650). Such was his passion for Aquinas's theology that when he went to live in Holland in 1629 he took few books with him other than his Bible and the works of St. Thomas (Russell, 1969:544). A life-long practising Catholic he shared Aquinas's love of reason and in the space of his own lifetime became renowned as a philosopher, mathematician and scientist.

Descartes' love of reason is communicated in his 'Discourse on Method' (1637) and particularly in his concept of 'Cartesian doubt' which he introduced in this work. Determining to doubt everything that is open to doubt, his intention was thereby to discover that which is beyond doubt and which is therefore by definition true. The result of this exercise, supposedly carried out in an oven (Russell, 1969:543), was his dictum 'I think therefore I am' as the basis of all that is knowable and as the reason for acknowledging the existence of the human soul. In presenting the process by which he reached this conclusion, Russell (1969:548) wrote 'I am a thing that thinks, a substance of which the whole nature or essence consists in thinking, and which needs no place or material thing for its existence. The soul, therefore, is wholly distinct from the body and easier to know than the body; it would be what it is even if there were no body'. From this Descartes postulated that, as consciousness can only have its origins in the realm of the spiritual, it must follow that it can only be identified with the soul. As he asserts that the soul is unique to humans he is left with only one conclusion regarding the status of animals. Animals have no soul, therefore they must be without consciousness – thus they must be mere automata.

As a result, whilst Russell (1969:542) is able to describe Descartes as 'the father of modern philosophy' and suggests that 'there is a freshness about his work that is not to be found in any previous philosopher since Plato', Singer writes of him in less flattering terms. Although he is ready to acknowledge Descartes' status as one of the world's greatest philosophers, Singer (1994:200) describes his views as 'the last, most bizarre, and - for the animals - most painful outcome of Christian doctrines.'

Singer's judgement is based on the obvious implications of Descartes' conclusions for the human/animal relationship. In relegating other species to the level of 'mere automata', Descartes was effectively releasing the human conscience from any sense of guilt over the abuse of animals. Furthermore, in a new scientific age with no recourse to anaesthetics, he was practically granting humanity a licence to use animals for live dissection. An acceptance of the Cartesian method with its mechanistic conclusions freed many scientists, including Descartes himself, from any qualms that they may have had about dissection, whilst consequently imposing untold suffering on the subjects of their experiments.

As Singer (1994:202) has commented, the reaction of many of Descartes' contemporaries to this abuse of animals was one of censure and horror. Voltaire for example described his own reaction, when he wrote in 'Dictionnaire Philosophique' of the 'barbarians who seize this dog, who so greatly surpasses man in fidelity and friendship, and nail him down to a table and dissect him alive, to show you the mesaraic veins! You discover in him all the same organs of feeling as in yourself. Answer me, mechanist, has nature arranged all the

springs of feeling in this animal to the end that he might not feel?'
(Singer, 1994:202).

The physiological similarities led to the conclusion that other
mammalian species might also share consciousness or emotions akin
to those experienced by humans. From this it was a simple step to
then suggest that they should therefore command our concern and
compassion. The Scottish philosopher David Hume (1711-76)
suggested, for example, that we are 'bound by the laws of humanity to
give gentle usage to these creatures' (quoted by Singer, 1994:202). In
a similar vein, writing in 'The Guardian' of May 21st 1713 Alexander
Pope objected to live dissections on the grounds 'that "although the
inferior creation" has been "submitted to our power" we are
answerable for the "mismanagement" of it.' Others giving voice to
the criticism of Descartes' mechanistic philosophy and its painful
implications for animals included the vegetarians Thomas Tyron
(1634-1703) and John Oswald (d.1793), and the writers Pupendorf,
Gerber and Winkler in Germany and Hildrop and Primatt in England.
Each one supported their censure by claiming that humans have a
moral obligation to care for other creatures and in doing so laid the
foundations for the eventual emergence of the animal welfare
movement.

Amongst the eighteenth century arguments against the mistreatment of
animals, however, one of the most influential focused on the
perpetrator of the act rather than the victim. Originally proposed as
early as the first century AD by Tertullian (c.155-222), and prior to
that by Plutarch, it was argued that the people acting cruelly towards
animals could be so brutalised by their actions that they might
eventually behave in a similar way towards those around them. It
seems likely that this argument first re-emerged in the eighteenth
century in a sermon on this theme published by James Grainger in
1772. Eight years later it appeared again in a different form in one of
Immanuel Kant's lectures to his students. Although Kant denied that
animals have any claim to human care and regard, he did believe that
humanity has an indirect moral duty towards them. This belief arose
out of his argument that our attitude and behaviour towards other
species reflects our attitude and behaviour towards each other. This
was expressed in his 'Duties to Animals and Spirits' in which he

suggested that 'so far as animals are concerned we have no direct duties. Animals are not self-conscious and are there merely as a means to an end. That end is man ... Our duties to animals are merely indirect duties to mankind ... If he is not to stifle his human feelings, he must practice kindness towards animals, for he who is cruel to animals becomes hard also in his dealings with men ... (whereas) ... tender feelings towards dumb animals develop humane feelings towards mankind' (quoted by Regan, 1988:178).

Whilst Kant's conclusions result in exhortations to show kindness towards animals, the motive behind that kindness leaves many questions unanswered. Indeed, it was these questions that were raised that same year by Jeremy Bentham (1780) in his 'Introduction to the Principles of Morals and Legislation' in which he established the criteria for many of the future arguments regarding the welfare of animals. In this he stated that 'the day may come when the rest of the animal kingdom may acquire those rights which never could have been witholden from them but by the hand of tyranny. The French have already discovered that the blackness of skin is no reason why a human being should be abandoned without redress to the caprice of a tormentor. It may one day come to be recognised that the number of legs, the villosity of the skin, or the termination of the *os sacrum*, are reasons equally insufficient for abandoning a sensitive being to the same fate. What else is it that should trace the insuperable line? Is it the faculty of reason, or perhaps the faculty of discourse? But a full grown horse or dog is beyond comparison a more rational as well as more conversable animal, than an infant of a day or a week, or even a month, old. But suppose they were otherwise, what would it avail? The question is not, "Can they reason?" nor "Can they talk?" but, **"Can they suffer?"'** (quoted by Singer, 1995:95). The ability of a creature to experience pain was now proposed as the sole criterion required for the condemnation of cruelty.

It was largely as a result of his applying the concept of rights to animals, and the formulating of his subsequent dictum, that Bentham has been accorded his status as the pivotal figure in the animal rights movement. It would actually be more accurate, however, to accord this position to Humphry Primatt who, in 1776, wrote 'The Duty of Mercy'. In this, he described how pain is a common experience of evil endured by both humans and animals. 'Pain is pain, whether it be inflicted on man or beast; and the creature that suffers it whether man or beast, being sensible of it whilst it lasts, suffers evil' (cited by Linzey, 1994:16). He also appreciated that the superiority enjoyed by humans over other species does not signify a right to abuse those species. 'Now, if among men, the differences of their powers of the mind, of their complexion, stature, and accidents of fortune, do not give one man a right to abuse or insult any other man on account of these differences; for the same reason, a man can have no natural right to abuse and torment a beast, merely because a beast has not the mental powers of a man' (quoted by Linzey, 1994 :16).

Sadly, Primatt's work has often been ignored by those concerned with animal welfare. Whilst, for example, both Singer in 'Animal Liberation' (1995) and Regan in 'The Case for Animal Rights' (1988) acknowledge their debt to Bentham, neither of them mention Primatt. It was his philosophy, however, that laid the foundation that they and many others have since built upon. As Brooman and Legge (1997:5) suggest, 'Primatt's concern for animals and his recognition of pain as a significant interest to be protected against in animals as well as ourselves presaged many of the arguments used by the animal rights lobby today.'

This growing appreciation of the commonalities shared by animals and humans received added momentum in the latter part of the nineteenth century with the introduction of Darwin's theories concerning the origin of the species. He 'explained that there are evident similarities between people and animals not only in terms of physical characteristics such as muscles, bones and reproductive systems, but also in terms of 'higher mental powers' such as memory, curiosity, imitation and reason which are seen in animals to varying degrees' (Brooman and Legge, 1997:15). Darwin's scientific approach to the whole subject of the man/animal relationship radically altered the way in which that relationship was perceived. As he emphasised the common links between the species, so he eroded many of the previous assumptions concerning the uniqueness of humanity and with them the justification for the mistreatment of animals. No longer could they be treated as mere automatons. In 'The Expression of the Emotions in Man and Animals' published in 1872, Darwin suggests that the evidence is to the contrary. Far from being unfeeling machines, the body language and facial expressions of animals would indicate that they too experience emotions. Whilst this view is dismissed as anthropomorphism by many of Darwin's critics it does have its supporters, including Masson and McCarthy, who in 1996 published 'When Elephants Weep: The Emotional Life of Animals'.

The potential implications of this new understanding of human responsibility towards other species were soon realised as pressure began to be applied to national governments across Europe to legislate against cruelty (see Section 3.4). At the same time, it also gave rise to a greater sensitivity towards animals amongst those who worked with them. The origins of traditionally accepted principles of stockmanship have already been discussed, but as the eighteenth century moved into the nineteenth, so there developed a general deepening awareness of human responsibility towards both other people and other species.

This new humanitarianism was manifested in legislative measures, which were to lead to many improvements for man and animal alike. Words from the eighteenth century German philosopher Wilhelm Dietler express something of this as regards animals, in suggesting that they should be 'slaughtered as quickly and as painlessly as possible. Cattle should be provided with adequate nourishment, domestic animals should be given a home and sufficient care, and animals used for work and transport should not be over-exerted except in emergencies' (cited by Maehle, 1994:85). Dietler's concern for the needs of livestock and the corresponding responsibilities of those caring for them was characteristic of the revolution in agricultural practice and especially livestock farming that was taking place during much of the eighteenth and nineteenth centuries.

3.3 The Agricultural Revolution

The Agricultural Revolution effectively began with the Enclosure Movement in the 16th and 17th centuries. The enclosure of open fields and commons by the planting of hedges changed the face of the English landscape and its agriculture forever. Without these enclosures, however, much of the work of livestock improvement, which characterised these and subsequent centuries, could never have taken place. Whilst the disastrous social consequences of enclosure have to be recognised, these agricultural benefits must also be taken into account. The social cost is summarised by Bonham-Carter (1971:39), who writes: 'For many smallholders, and for most cottagers and squatters, it meant actual loss of living and independence; and there quickly arose a large class of landless labourers.' Likewise, George Ewart Evans (1956:253) draws attention to the injustices that many small farmers suffered as a result of enclosure and quotes lines from an old rhyme on the theme:

'They hang the man and flog the woman
Who steal the goose from off the Common;
But let the greater criminal loose
Who steals the Common from the goose.'

Rather than simply passing judgement on those responsible, however, Bonham-Carter recognises that such measures were as unavoidable as they were ultimately beneficial. 'Looked at simply as an historical event, the Great Enclosures appear inevitable. They enabled millions of acres of poor grazing, indifferent arable and wasteland to become productive. Without this addition to tillage, under the pressure of population and war, this country would have neared starvation long before Waterloo. Moreover, enclosure benefited agriculture as a whole' (Bonham-Carter, 1971:39). This view is in accord with those benefits to livestock production, which have already been described in Section 3.1 as consequences of the widespread enclosure movement. As has been noted, such agricultural improvements included controlled breeding, disease prevention and the development of animal husbandry as a science, all of which were to lead to untold benefits for livestock and producer, as well as consumer. In addition, they were to lead to the greatest progress in agricultural practice and knowledge that man had ever known. Seebohm (1976:262), in referring to eighteenth century farming, states that 'during this century there spread over England the greatest advance in the practice of farming that had taken place since agriculture was first established in this country.'

These improvements were very much in tune with the general spirit of eighteenth and nineteenth century England. It was the age of industrial expansion, engineering expertise, scientific discovery and philanthropic endeavour, and all of these were to have an influence on the way in which agriculture was to develop. Engineering, for example, was to transform the world of the arable farmer. Beginning with the very basic invention of the seed drill described by Jethro Tull in 1731 in 'The New Horse-hoeing Husbandry', it eventually led to the introduction of mechanised power onto the land with the first application of the steam engine for agricultural purposes. It also had an impact on livestock production. Although the first example of the direct application of engineering to livestock husbandry was probably the introduction of milking machines in the late nineteenth century, engineering did have an influence on the livestock industry at a much earlier date. The engineers, for example, who drained the Fens in the seventeenth century, not only opened up new acres for arable crop production but also made available new possibilities for livestock producers in terms of the potential for new grass pastures on highly productive soil. During the eighteenth century, there was increased use of fenland for the fattening of cattle and sheep that were driven from Scotland and Wales for eventual sale in London, a practice that would have been impossible were it not for the work of the drainers (Thirsk, 1957:207). As Watson and Hobbs (1951:22) comment, drainage made 'possible the full realisation of all the discoveries and inventions which followed,' and regarding stock rearing, 'the greater stock-carrying capacity of the drained lands led to a general improvement of breeds.'

Prior to the draining of the Fens, animals kept on them had to endure much hardship. Evidence of this has survived in the works of the eighteenth century agricultural commentator Arthur Young. Describing the condition of stock kept on Wildmoor Fen before it was drained, he writes, 'In winter horses were driven to such distress for food that they ate every remaining dead thistle and were said to devour the hair off the manes and tails of each other. In one year over 4,000 sheep rotted, and even the young geese died of the cramp on the water-logged commons' (cited by Watson and Hobbs, 1951:33). Young's critical observations of old agricultural practices, which resulted in such suffering, fuelled his passion for new ideas. Many of these are contained in 'The Farming Kalendar', which he published in 1771, although sadly, like Tusser before him, he was more successful in putting his ideas on paper than on making them work in practice. As a writer, however, he was to prove a major influence in the world of agricultural improvement. He wrote over two hundred and fifty volumes on subjects related to farming and in 1784 published an agricultural periodical with the title 'Annals of Agriculture'. This included contributions from a number of important figures, including George III, who wrote under the pseudonym of Ralph Robinson.

Many of Young's views, although innovative at the time, are now obviously out-dated. For example, he discusses the merits of buying wether sheep at two or three years old for fattening, a practice that would attract few buyers in today's market! Likewise, his definition of good farming was based on different criteria than those that would apply now, as his chief concern was for the amount of manure produced! 'Respecting the quantity; therein lies the proof of your being a good farmer; perhaps the most important, convincing proof that a farm can offer' (Young, 1771:132). Such a measure of good husbandry is hardly appropriate today.

Many of his other observations are still relevant, however, especially with regard to the responsibilities and standards of stockmanship. Writing for shepherds, and advising them about their responsibilities to their flocks, he states that their 'business throughout the year, on whatever food, is to keep them (their sheep) in good and healthy order.' Shepherds are to confine their sheep 'at night in a sheep yard, well and regularly littered with straw, stubble or fern; by which means you keep your flock quite warm and healthy in bad seasons' (Young, 1771:7). He advises those who keep cattle to 'take care that they are regularly supplied with straw and that they always have water at command' (Young, 1771:8). The yards are to be generously littered 'so that cattle may always lay perfectly dry and clean. Their health requires this attention' (Young, 1771:9).

Young's overall concern is for profitable agriculture, but it is clear that he believes that this requires a combination of good husbandry in terms of applied practical knowledge, which is allied to skilled stockmanship. This he equates with a caring, responsible attitude towards the animals over which one has charge, a principle that he considers to be an integral part of profitable and responsible farming.

Young recognised that the integrity of agriculture was dependent on a balanced approach to production. For example, the reason he exhorted farmers to produce as much manure as possible from their stock, was that this would feed the land and improve its fertility. Thus, the success and profitability of arable farming depended on well-managed livestock production. Similarly, the livestock producer was dependent on arable crop production for food for his animals. Agriculture was shaped by a sense of interdependence between the different sectors, and this in turn must have made for a more holistic attitude towards the industry as a whole. In effect, many aspects of the modern concept of agricultural sustainability have their roots in the work of the improvers of the eighteenth and nineteenth centuries.

It is worth noting, however, that many of the current definitions of agricultural sustainability are notoriously vague, as is illustrated by Aaker (1994:2). He cites two examples. The first comes from Olson, described as 'an American farmer and pastor' who defines it as 'an agricultural system that nurtures and beautifies the land, while providing health, abundant food, and a good living for farming families'. The second from Gips of the International Alliance for Sustainable Agriculture, whose definition describes it as a system that is 'economically viable, ecologically sound, socially just and humane.' Consequently, Spedding (1994:1-8) suggests that such definitions have become so generalised that they are virtually meaningless. As a result he offers a 'package' of six 'sustainability attributes' for animal production systems that provide a more specific definition of the concept. He contends that 'they should be highly productive, of safe, high quality products ... physically sustainable ... biologically sustainable... should satisfy agreed criteria for human and animal welfare, ...must not give rise to unacceptable pollution, by-products or effects... and they must be profitable' (Spedding 1994:7). With his emphasis on high standards of animal welfare, the need for profitability and the interdependence of physical and biological resources, he echoes the principles expounded by the eighteenth century improvers.

Thomas Coke exemplifies many of the attributes of these improvers. Having succeeded to the Holkham estate in 1776 he personally managed the estate farms and began to apply some of the new ideas *vis-à-vis* the new farming methods. He worked at improving the land and improving the stock, and at the same time encouraged his tenants to follow his example. He realised, however, that if he was to be successful in achieving this latter aim then he had to provide those tenants with an incentive. This took the form of extended leases, which not only gave greater security and confidence to the tenants but also led to a more stable farming community. The combination of these two features of life on the Holkham estate gave Coke the very results he desired. His tenants were more amenable to his new ideas and suggestions, whilst the stability he had created resulted in the very agricultural environment in which those methods could work. With regard to livestock, those methods included a concentration on breed improvement, the application of new ideas concerning the nutritional needs of the animals and the ways in which those needs could be met.

In the seventeenth century, Sir Richard Welton had returned from Holland enthused by Dutch ideas concerning crop rotation and the subsequent use of turnips as winter stock feed. In c.1645-50 he published 'A Discours of Husbandrie used in Brabant and Flanders' extolling these ideas (Fussell, 1969:35-40). The implications for the health and well-being of over-wintered animals were tremendous, but despite Welton's efforts it was only when Viscount Townshend began to promote these ideas that they began to catch on. Described as the 'prototype of the new aristocratic squires' (Bonham-Carter, 1971:35) Townshend was to be a leading figure amongst agricultural improvers, and although there were others, even before Welton, who attempted to introduce crop rotation in England, it is he who is credited with establishing its widespread use (Bonham-Carter, 1971:34). His four-course system of wheat, turnips, barley, and clover and rye grass, produced increased supplies of available forage and thus improved conditions for livestock, especially during winter, making them healthier and better-fed than ever before.

What Townshend achieved for stock nutrition, others undoubtedly achieved for stockbreeding. Bakewell for example has been identified as the 'founding father of British stockbreeding' (Bonham-Carter, 1971:37). This reputation was the result of his producing the New Leicester, which was to establish the criteria by which all subsequent British breeds of sheep were to be judged, and which was to become the forerunner of the Border Leicester. Working with other bloodlines, John Ellman of Gynde in Sussex, concentrated on breeding from the best stock he could find. Using only sheep of a particular type and concentrating on carcase quality, he eventually produced the Southdown breed, which 'became and remains to this day the ideal butchers' sheep, with minimum waste, and maximum development of the cuts consumers prefer' (Fraser, 1954:84).

Innovations in sheep breeding were paralleled by similar developments in the world of cattle rearing. Bakewell's ideas and methods, for example, were introduced to the Colling brothers of Darlington by a mutual friend, Robert Culley, and as a consequence they began to develop the Shorthorn type cattle. These were described as 'an animal of moderate size, shapely and early maturing, with well-sprung ribs, short legs and good thick hide covered with thick mossy hair' (Fream, 1973:593). Two Yorkshire breeders, Thomas Bates and Thomas Booth, developed the breed further, with Booth concentrating on the beef characteristics and Bates on the breed's potential as a dual-purpose animal for beef and milk production. In 1671, the traditional Hereford cattle had been crossbred with white faced Flemish beasts, producing the forerunner of the modern Hereford breed. It was largely as a result of the work Benjamin Tomkins and his family, however, that the Hereford as it is now known was developed between the years 1742 and 1815 (Fream, 1973:593).

Similar stories of breed improvement and development are found throughout British agriculture during the eighteenth and nineteenth centuries. Hugh Watson of Keillor in Forfarshire, for example, established the Aberdeen Angus breed in the early 1800s as a result of applying the breeding principles used on the Shorthorns. In the 1820s the first Channel Island cattle were imported into England. Herds were set up in Cornwall and the Isle of Wight, and the principle of breeding purely for dairy production rather than aiming for a dual-purpose beef/dairy animal was firmly established (Fream, 1973:598-600).

By the early nineteenth century, various principles and systems of livestock production were established that were to set the pattern for all that was to follow. Engineering, business acumen, and science all played their part with the latter epitomised for farming in John Benet

Lawes' founding of the first Agricultural Experimental Station at Rothamstead in 1834 and the establishing of the Royal Agricultural Society in 1838 (Fussell, 1969:56). The Society received its Royal Charter in 1840 and its motto 'Practice with Science' well typifies the spirit of agriculture at this time. This approach to farming was further encouraged by the many businessmen who had made their fortunes in the Industrial Revolution and who now used that money to buy country estates. Familiar with new ideas and innovative solutions to the problems of production, they were quick to follow the example of the agricultural improvers and began to apply the lessons they had learnt in the world of manufacturing industries on their estate farms.

The net effect of all of the improvements in stock rearing, handling, and nutrition was obviously beneficial for farmed animals, especially when combined with additional developments in veterinary science. Whilst the Royal Veterinary College was founded as early as 1791, it was some time before the resulting changes in veterinary medicine and practice impacted upon the farming community (Seebohm, 1976:333). During the nineteenth century, however, there were major advances in medical and veterinary science, which resulted in a wider understanding of the needs, treatment and medication of sick animals. In 1858, for example, Virchow published a book on cellular pathology, which introduced the concept of the cell as the centre of all pathological change (Naidoo and Wills, 1994:7). In 1864, Pasteur first isolated organisms under the microscope and, in 1882, Koch managed to isolate the tubercular bacillus (Naidoo and Wills, 1994:7).

Although the moves throughout this period towards ever-higher standards of livestock production and care emanated from a revolution in agricultural and scientific methods, they were more than just the incidental side effects of a wider process. Concern for the well-being of animals was in fact one of a number of manifestations of the developing social conscience of Victorian England. This conscience was to bring about major social and moral changes in the human/animal relationship as growing pressure was brought to bear on the British government to legislate against cruelty to animals and to make such cruelty a criminal offence.

3.4 Social and legislative developments

Part of the English cultural mythology is that we are a nation of animal lovers. The evidence would suggest otherwise. Ritvo (1994:107) quotes Queen Victoria's observation from 1868 as to how 'the English are inclined to be more cruel to animals than some other civilised nations are' whilst a journalist described England in 1825 as 'the Hell of dumb animals' (Ritvo, 1994:107). Similarly, a writer in an edition of 'The Zoologist' from 1853 complained of 'the continual and wanton persecution of birds (and animals) in England, as contrasted with their kind treatment in Norway' (Ritvo, 1994:107).

The accuracy of this judgement is confirmed by the public humiliation and scorn suffered by those who first attempted to legislate against animal cruelty in the early 1800s. Indeed, the would-be reformers were subject to widespread ridicule and opposition from both the political establishment and the general public. 'The Times' for example considered it an issue 'beneath the dignity of Parliament' and George Canning, a future Member of Parliament, defended bull-baiting on the grounds that 'the amusement inspired courage and produced a nobleness of sentiment and elevation of mind' (Ritvo, 1987:125). Such views were commonly held, for whilst intellectuals engaged in academic debate about the status of other species, the majority of the population continued to accept cruelty to animals as a fact of life and as a popular source of entertainment. This is hardly surprising when one takes into account the harsh conditions that most people had to endure. Hardened by their own circumstances they were unlikely to feel a great deal of sympathy for the suffering of other creatures. In fact, they found pleasure in witnessing that suffering.

It was because the cruel spectacle of the cockpit and the baiting ring aroused such violent emotions amongst the spectators, that fear of what might result brought about calls for the banning of these sports. As Ritvo (1987:131) suggests, the opposition to baiting was not born out of the English love for animals, but was rather motivated by a fear of the ill-disciplined crowds that attended such gatherings. The law was directed more at protecting society from riots than protecting animals from cruelty. The general acceptance of this anthropocentric attitude towards animal welfare can be illustrated from both philosophy and law. The issue in question in matters concerning cruelty to animals tended to be harm to the person, either morally or financially, rather than harm to the animal.

As has already been noted, there was a strong anthropocentric element in Kant's teaching, which was to have a wide influence on public attitudes and perceptions regarding the treatment of animals. Cruelty was condemned, not because of the harm done to the victim but because of the brutalising effect that such cruelty had on the perpetrator. This attitude was succinctly expressed in a sermon published in 1772 by James Grainger in which it was stated that 'callousness against animals is training for cruelty against humans' (quoted by Maehle, 1994:84).

The law reflected a similar emphasis in that animals were deemed to be property. This meant that, in any legal case involving cruelty against an animal, judgement was based solely on the consequential financial harm suffered by the owner. No legal judgement was possible against someone accused of cruelty unless it could be demonstrated that their actions directly harmed the interest of the owner of the animals concerned. Effectively, whilst legislating against a third party who inflicted suffering on an animal, the law allowed an owner a free hand in how he treated his 'property'. As Brooman and Legge (1997:31) have noted, 'for centuries animals had no legal protection, being treated merely as property, not deserving of welfare or protection.' Also, 'although extremely valuable to the progress of humanity the treatment of animals as property has, arguably, been extremely damaging to the fate of animals. As property, animals' owners have been able to dispose of their animals as they wish' (Brooman and Legge, 1997:53). In addition, despite their being treated as property, there were long-established laws in many countries, which deemed animals morally culpable. Consequently, an animal could be prosecuted for any crime and many were not only tried but were also punished. The last known occasion when this occurred was as recently as 1906 when a dog was sentenced to death in Switzerland (Brooman and Legge, 1997:37).

In many European countries, however, the eighteenth and nineteenth centuries saw the emergence of a more humane attitude towards animals. Although there had been no legislation prior to the nineteenth century that actually protected them, there are accounts of previous court proceedings in which those responsible for inflicting suffering on animals were condemned. There is an account, for example, of a trial in Prussia in 1684 when a man was fined and imprisoned in the pillory for two days for mistreating his horse (Maehle, 1994:95). Similarly, sixteenth century Berlin had an ordinance punishing anyone guilty of torturing an animal with exposure in the pillory (Maehle, 1994:95). It was not until the nineteenth century, however, that any major advances were made with regard to the legal protection of animals. This is illustrated again from Germany, where some of the first Acts were passed, notably in Wurtemberg, Hanover and Prussia in 1839, 1847 and 1851 respectively (Maehle, 1994:98).

There is similar evidence from eighteenth and nineteenth century England of a growing concern for the welfare of animals. In 1798 for example, Thomas Young published his 'Essay on Humanity to Animals' in which he demanded that animals be given protection under the law. This was followed in 1800 by the placing of a Bill before Parliament demanding an end to bull-baiting. Introduced by Sir William Pulteney it was narrowly defeated by 43 votes to 41 (Brooman and Legge, 1997:40). Reintroduced by Wilberforce in 1802, it was again defeated. The advocates of animal welfare presented a further Bill in 1809 aimed at protecting domesticated animals, especially horses and cattle. Introduced by Lord Erskine, it was also defeated, as was another attempt at legal protection for horses, which was put before Parliament by Richard Martin in 1821.

Martin, dubbed 'Humanity Dick' by George IV, re-presented the Bill the following year and this time was successful. The 'Ill Treatment of Horses and Cattle Bill', known as Martin's s Act, made it an offence 'to beat abuse, or ill-treat any horse, mare, gelding, mule, ass, ox, cow, heifer, steer, sheep or other cattle' (Brooman and Legge, 1997:46). The penalty was a minimum fine of ten shillings and a maximum sentence of two months in prison. 'Humanity Dick' so believed in the Act and the principles that lay behind it that he actually brought the first case to court. Samuel Clarke and David Hyde were accused and subsequently found guilty of beating horses at Smithfield Market and were fined twenty shillings each (Brooman and Legge, 1997:46).

In 1825, Parliament agreed to outlaw bull-baiting and this was followed in 1835 with an Act to extend the scope of the 1822 provisions. This was proposed as an Act to 'consolidate and amend several Laws relating to the cruel and improper treatment of animals, and the Mischiefs arising from the driving of Cattle, and to make other provisions in regard thereto' (Brooman and Legge, 1997:46).

It was as if the move to protect animals had gained a momentum of its own, for in 1849 the 'Cruelty to Animals Act' was passed strengthening the powers of the Law even further with regard to animal welfare. One effect of this legislation was that it was now possible to prosecute not only an employee who was cruel to stock, but also his employer. The repercussions for agriculture were obvious, for the employer was legally liable for the way his workforce treated stock and it was in his interest to ensure that they aimed for the highest standards of animal welfare. The Act had a further impact on farming - for the first time it drew attention to the welfare of animals going to slaughter and required that they be fed and watered.

It is important to recognise that in all these moves to consolidate animal welfare measures the guiding principle accepted by Parliament was that 'while man is free to subjugate animals, it is wrong for him to cause them suffer unnecessarily' (Brambell, 1965:7). This remains the foundation of all English legislation regarding animal welfare.

Although a number of further laws regarding animal welfare were passed during the following decades, few of these were concerned with agriculture. For the farming community, the next major step regarding legislation that was relevant to livestock production came in 1911 with the Protection of Animals Act. This effectively consolidated most of the previous measures, to the extent that the 1911 Act is now considered to be the basis of all English legislation regarding the legal status of animals. It made it an offence 'cruelly to beat, kick, ill-treat, infuriate or terrify any animal... or wantonly or unreasonably by doing or omitting to do any act, cause any animal any unnecessary suffering' (Blackman, Humphreys, and Todd, 1989:101).

The 1911 Act also continued to uphold the precedence of human interests over those of animals and to retain their legal status as 'property'. Both views are unpopular with the advocates of the philosophy of animal rights, but as they describe the status and definition of animals under English Law, they also form the legal framework in which livestock farming operates. As such they clearly have an influence on what is considered to be morally acceptable *vis-a--vis* husbandry systems and practices and have an overall effect on the attitude of those engaged in livestock production.

Whilst the original impetus regarding animal welfare legislation was largely motivated by concern for draught animals or those used for the purposes of human entertainment many of the post-1911 Acts regarding animals were of benefit to farm livestock. The Protection of Animals (Anaesthetics) Act for example, passed in 1954, forbade the performance of specific surgery on animals without anaesthetics. This one legal measure ushered in a new era in livestock husbandry and stockmanship, as it paved the way for the outlawing of many painful procedures such as the unanaesthetised debudding of cattle horns. It is important to recognise, however, that the legislative process has not been the only expression of this concern for animals that has been such a feature of English life since the end of the eighteenth century. Other factors have also influenced the general attitude towards the use and treatment of stock. Some of these were touched on by Maehle (1994:81) who, amongst other things, referred to changing moral perceptions, the wider acceptance of vegetarianism and a greater understanding of animal psychology, all of which will be considered more fully in Chapter 4.

One of the major influences over the past two centuries, however, has been the birth and subsequent flourishing of the animal welfare movement. The latter part of the twentieth century has witnessed the burgeoning growth of groups dedicated to the relief of animal suffering. Frequently focusing on specific issues, their opposition has been directed against such practices as live animal exports, the keeping of wild animals in zoos, hunting with hounds and intensive livestock farming. Their origins, however, are back in the early 1800s, when there were a number of attempts across Europe to establish societies aimed at promoting the welfare of animals.
The first efforts in Liverpool in 1809 and London in 1822 proved unsuccessful. Success was achieved on 16th June 1824, however,

when 'A Society instituted for the purpose of Preventing Cruelty to Animals' (SPCA) was established. This was largely due to the efforts of the Reverend Arthur Broom who first floated the idea in an advertisement in the magazine 'John Bull' in November 1823. In their brief biography of Broom, the RSPCA (1995) describe how he was so influenced by Primatt's 'A Dissertation on the Duty of Mercy and the Sin of Cruelty to Brute Animals' that he personally produced the second edition in 1823 with all profits being donated to the SPCA.

In 1824 Broom decided to devote himself totally to the work of the Society and to this end resigned his living of St. Mary's Bromley, St. Leonard. Unfortunately, the finances of the SPCA were precarious and, with no income of his own, he was sent to prison for a short time in 1826 and was only released with the help of Richard Martin and Lewis Gompertz. Two years later Broom resigned as the Society's secretary and was succeeded by Gompertz. Although Broom continued on the Committee until 1833, his diminishing role in the work of the SPCA was such that his death from consumption in 1837 received no mention in their records. Despite all that he personally achieved for the cause of animal welfare, it is a tragedy that Broom's own life ended inauspiciously with a pauper's funeral in Birmingham cathedral, and burial in an unmarked grave (RSPCA, 1995).

The SPCA in contrast grew in strength and influence. Granted royal patronage by Queen Victoria it became the Royal Society for the Prevention of Cruelty to Animals in 1840 and as such was to set the standard for much of what was to follow both legally and ethically. Aimed at changing human attitudes as well as the conditions that many animals have to endure, its work remains much the same today as in the 1800s. This is illustrated in the original Prospectus, prepared in 1825, which states that the purpose of the Society was to be:

1. The circulation of suitable publications, distributed gratuitously, or at a low price, among the less enlightened orders of the community, particularly among persons entrusted with the care of cattle, such as drovers, coachmen, waggoners, etc.

2. The introduction of works calculated to make salutary impression on the minds of youth, into our schools of education.

3. The establishment of periodical discourses, from the pulpit, in one or more metropolitan churches.
4. The employment of Inspectors, in the markets and streets; and the giving of publicity to cases that may occur of particular atrocity.
5. Frequent appeals to public feeling, through the medium of the press, in every practicable mode which may tend to awaken a more general attention to this important and interesting project.' (Source : RSPCA, 1995)

The above aims remain much the same today, as is indicated by a more recent statement: 'The Society achieves change through a careful balance of powerful campaigns directed towards both government and consumers, scientific research and liaison with the food and farming industry to promote the development of high welfare farming systems' (RSPCA 1995). Whilst the activities of the Society are concerned with all aspects of human/animals interaction, a major part of their work is focused on welfare issues, which arise out of livestock production and husbandry systems. In stating their position regarding the farming of animals the RSPCA express their opposition 'to all farming systems which cause distress of suffering or deprive animals of the opportunity to indulge in their natural behaviour' (RSPCA, 1995). On the positive side their 'Freedom Foods Campaign' which will be discussed later in Chapter 5, is a practical expression of their commitment to the establishing of more welfare-friendly farming systems in the UK.

3.5 Continuing change in agriculture and animal welfare

The major developments in agricultural improvement that took place in the eighteenth and nineteenth centuries were largely fuelled by the burgeoning population of the industrial towns and cities. The population of 5.5 million in 1700 had grown to approximately 9 million by 1800 and to 36 million by 1900 (Richards and Hunt, 1965:viii), consequently bringing about a dramatic increase in the national demand for agricultural produce. The improvements in livestock breeding (see Section 3.4), resulted in an increase in the production of meat, milk and eggs, but for the demand to be fully met there had to be parallel changes in production systems. Without these developments, large numbers of the Victorian poor would have been reduced to starvation and famine.

An example of the moves towards new production systems can be found in the developments that took place in the dairy industry at the end of the nineteenth century. These effectively began with the introduction of the milking machine in the 1890s. Originating in Denmark, the first successful system to be launched in Britain was the Thistle, which was introduced in 1895 (Fussell, 1969:64). Although it was well into the twentieth century before their use became universal, those early machines set in motion an inevitable progression towards ever more efficient systems of milk production. By the 1920s, that efficiency was further improved with the introduction of new methods for feeding dairy herds. The work of MacIntosh at Reading, and Boutflower at both Harper Adams College and the Royal Agricultural College in Cirencester spearheaded a new understanding of the nutritional needs of the dairy cow (Watson and Hobbs, 1951:172). This was to have dramatic results not only for herd productivity but also for the overall health and welfare of the individual cows.

Subsequent advances in breed development, nutrition, veterinary science, and the application of technology have turned the dairy cow into a highly specialised creature designed and bred purely for milk production. As a result, the older dual-purpose breeds such as the Dairy Shorthorn so favoured by nineteenth century farmers have now largely disappeared from commercial dairy units in favour of the Friesian and the Holstein. At the same time there has been a sharp decrease in the number of dairy farms in the UK, accompanied by a consequential growth in the herd size and in milk production per cow (see Table 1).

Such developments have in turn raised a plethora of ethical and welfare issues. These range from the disposal of unwanted bull calves to the over-stressing of cows and the use of questionable animal feeds, all of which will be discussed in future chapters.

A major factor in this drive towards ever-greater efficiency and output was the need to feed a beleaguered nation during the Second World

War. Farmers were put under great pressure to produce as much food as possible as efficiently as possible and those who failed to achieve the required standards had their land taken from them by the County War Agricultural Executive Committees (Bonham-Carter, 1971:87). Such an incentive was bound to have an effect in driving agriculture towards the goal of maximum production. This emphasis is further reflected in the measures set out in the Agricultural Act of 1947. Against the back-cloth of war-time food crises the purpose of the Act was 'to promote a stable and efficient industry capable of producing such part of the nation's food as in the national interest it is desirable to produce in the United Kingdom and to produce it at minimum prices consistent with proper remuneration and living conditions for farmers and workers in agriculture and with an adequate return for capital invested.' (HMSO, 1947 cited by Hill, 1975:61)

Table 1. The structure of the UK dairy industry:
1970 and 1995 (FAWC, 1997:4)

	1970	1995
No. of registered milk producers	100,741	36,709
Milk sales per producer (litres)	112,500	383,045
Average herd size (cows)	30	71
Average milk yield per cow (litres per lactation)	3,750	5,395

The combined aim of maximum production at minimum cost resulted in increased pressure being put on farmers to apply the methods and principles used in other industries to achieve ever-greater efficiency. Consequently, the changes in the dairy industry that have already been outlined are reflected in other areas of livestock production.

In the poultry sector, for example, one of the major and most controversial developments, which took place in the 1920s, was the introduction of the first battery cage systems from the United States. Interest was at first slow, but by the 1950s this pressure to achieve improved levels of efficiency and production had led to most eggs in the United Kingdom being produced in such systems. Writing in 1964

Harrison stated that 'Today, half the intensively housed laying birds are estimated to be in batteries, that is 40 per cent of all layers. In 1961, a 'poor' year it seems, there were seventy million laying birds in the United Kingdom, so there were some twenty-eight million birds in battery cages' (Harrison, 1964:40-41). Similar changes were taking place in pig production, with a post-war move towards specialised pig units in which 'industrial' type processes were applied and restrictions were placed on the movement and activities of the animals. These developments achieved their main goal, in that livestock production was greatly increased in most sectors in the three decades immediately following the 1947 Act, as is illustrated in Table 2.

Table 2.   Output of livestock commodities in the UK:
1950-1971 (adapted from Hill, 1975:27)

Commodity	Units	Pre-war Average	1950/ 51	1960/61	1970/71
Beef & veal	'000 tonnes	578	600	772	982
Mutton & lamb	'000 tonnes	195	147	241	231
Milk	million litres	7064	9080	11,100	12,503
Pork	'000 tonnes	171	59	441	619
Poultry meat	'000 tonnes	89	78	307	575
Hen eggs	millions	5158	6343	12,516	13,932

As in any industry the twin goals of efficiency and productivity in agriculture are driven by the need for that industry to be profitable. Without profit, the industry cannot exist. I suggest, however, that most people directly concerned with livestock production would add another factor to the equation, that of animal welfare. The principles of stockmanship that have already been outlined demand that a husbandry system whilst being efficient and profitable, must also be humane. It is not a case of any one goal having priority; in any civilised society they must be mutually dependent. 'We cannot abandon economics in favour of ecology or ethics; they have to co-exist and serve each other' (Britton, 1983:10).

It was not economics that was abandoned, however, but ethics, or so it was argued by Harrison who in 'Animal Machines' (1964) questioned the ethical acceptability of certain husbandry systems and popularised the term 'factory farming'. The publication of this book was to prove pivotal in the development of the animal welfare movement in the second half of the twentieth century. In fact it was as a direct result of Harrison's publicising the issues surrounding modern farming systems that the government established 'The Technical Committee to Enquire into the Welfare of Animals kept under Intensive Livestock Husbandry Systems' (Stewart, 1989:202). Now commonly known as the Brambell Committee, after its chairman Professor Roger Brambell, its terms of reference were: 'To examine the conditions in which livestock are kept under systems of intensive husbandry and to advise whether standards ought to be set in the interests of their welfare, and if so what they should be' (HMSO, 1965:1).

The standards and recommendations that the Brambell Commission eventually proposed were to set the benchmark for some of the most positive developments in UK farm animal welfare over the next forty years. As a direct response to its findings, for example, the government established the Farm Animal Welfare Advisory Committee. In 1979, this became the Farm Animal Welfare Council (FAWC), with direct responsibility for instituting the welfare standards for all farm livestock in the UK (see Chapter 5). The Commission's recommendations ranged from suggestions to the government on the improving of animal welfare legislation and the provision of education and training for stock persons, through to detailed technical advice to the industry in terms of husbandry systems (HMSO, 1965:60-65). Two further important points worth noting are that in making their recommendations the committee were concerned that:

- The competitiveness of producers should not be adversely affected and that if this happened then the government should redress the situation. This has implications in terms of current legislation on the phasing out of sow tethers and battery cages.

• The report should not be seen as a condemnation of British farmers for 'in our opinion the great majority of farmers are concerned to ensure the welfare and health of their stock' (HMSO, 1965:62).

The Committee debated the definition of animal suffering at length as they recognised that, whilst there are many difficulties in trying to understand how an animal experiences suffering, any recommendation without such a definition would be meaningless. The use of analogy, in terms of comparing their pain with that experienced by humans, was believed to be unavoidable. 'It is justifiable to assume that the sufferings of animals are not identical with those of human beings; it is equally justifiable to assume that they suffer in similar ways' (HMSO, 1965:9). The criteria decided upon and used throughout the enquiry was established by the earlier Report of the Departmental Committee on Experiments on Animals (The Littlewood Report), and it was this definition that was suggested as the basis for all future recommendations regarding animal welfare. It describes animal suffering as:

• 'Discomfort (such as may be characterised by such negative signs as poor condition, torpor, diminished appetite);
• stress (i.e. a condition of tension or anxiety predictable or readily explicable from environmental causes whether distinct from or including physical causes);
• pain, recognisable by more positive signs such as struggling, screaming or squealing, convulsions, severe palpitations' (HMSO, 1965:80).

What Harrison's book achieved in terms of making the public aware of ethical and welfare issues *vis-à-vis* intensive livestock production, the Brambell Report achieved in setting the parameters for the legislative and administrative oversight of such issues in agriculture. This in turn has established the criteria and pattern for developing and improving livestock production in the UK for the foreseeable future.

3.5.1 Economic change

Since FAWC was established in the 1970s, British agriculture has continued to undergo many changes. The period has been marked, for example, by major advances in technology and biotechnology, and dramatic developments in the marketing of farm produce. One of the most devastating changes however, has been the sharp decline in average farm income. In 1998 the NFU commented that 'net farm income measured in real terms has shown an underlying declining trend since we joined the EU in 1973' (NFU, 1998:34). A year later Deloitte and Touche Agriculture reported that their clients had suffered 'the worst results for a decade... net farm income plummeted by 54% to just £57 a hectare' (Deloitte and Touche, 1999:1).

The impact of this devastating decline was particularly felt in the livestock sector with dairy producers, for example, reporting that the average dairy farm income for 1999/2000 would be £8000. This amounted to a 20% reduction from the previous year and a 76% reduction from 1995/96 (NFU, 2000:3). Similar accounts were reported from across the livestock industry with pig farmers in particular reporting dramatic losses (NPA, 1999:1). As a direct result of this economic crisis many producers were forced to engage in activities that offended both their standards and beliefs regarding animal care.

This was illustrated in an article which appeared in 'The Independent' (Watson-Smyth 17th. April 99) and which reported how the RSPCA had had to accept a consignment of sheep from Welsh hill farmers who could neither afford to feed them nor pay the charges to have them destroyed. The organisation's spokesperson criticised the farmers' behaviour, stating that it was 'utterly irresponsible' and that it 'compromised the welfare of the animals.' For those men, however, it was their last resort as they described how they had nowhere else to turn and that they 'had not got the money to see the animals through the winter.' No matter how much they might have respected and cared about those animals, financial stringencies made it impossible for them to do so. Indeed, I would argue that the fact that they turned to the RSPCA for help was an indication of their concern for the welfare of their stock and that they didn't want to see them suffer.

This incident illustrates the tension that many producers have experienced. On the one hand they have wanted to rear their stock according to the ethical principles that undergird their industry and their own businesses, but on the other they have been unable to afford to do so.

Dairy farmers have been faced with such a dichotomy since the ending of the Calf Processing Aid Scheme in July 1999. As the public's confidence in the safety of beef was shaken in 1996 by the announcement of possible causal link between BSE in cattle and vCJD in humans, so the government introduced this scheme as a means of restricting the number of calves entering the food production chain for fattening. The price of beef had slumped, calves from dairy herds were valueless and as a result a mechanism was put in place which allowed farmers to be paid a set amount for newly born calves that were taken off the farm to be slaughtered.

Initially many dairy producers found it difficult on moral grounds to accept the scheme, as the treating of calves as the unwanted and disposable by-product of the dairy industry offended their strongly held beliefs in a stockman's responsibility towards all his stock. In reality, however, this is exactly what they were, and the economic pressures faced by dairy producers resulted in their having to come to terms with the necessity of disposing of calves in this way.

When the scheme came to an end, however, dairy breed bull calves were still virtually worthless and farmers were left with a yet more difficult choice that was in many ways similar to that of the sheep producers already mentioned. Rearing the calves to slaughter weight would be expensive and possibly pointless, in that the finished beasts would not only be worth little, but they would also bring poor quality carcases onto an already well-supplied market and thus reduce the value of the better animals from beef breeds. On the other hand, the delivery and slaughter costs incurred in having the young animals destroyed at an abattoir would be prohibitive. This effectively left producers with just one option, that of on-farm slaughter, with the young stock being killed as soon as possible after birth by a local huntsman, slaughterman or vet.

This unwanted development in the industry caused many producers to suffer both emotional and moral distress as they became directly involved not only in the slaughtering of young calves but also, by implication, in the young animals being treated as dairying's unwanted by-products. What had previously been done at a distance now became personal and therefore disturbing, even though they knew that the future of their livelihood, and therefore the rest of the herd, depended on it. As one dairy farmer commented in an article in 'The Independent' in the autumn of 1999 (Leach, undated) 'I can't afford to tag them or send them away to slaughter. I've had to shoot about 15, and there will be more. It's a desperate thing to do because my job is to rear livestock.'

The scale of the problem was indicated in 'The Farmers Weekly' (17.3.2000:8) report regarding the results of a survey conducted by the Farmers Union of Wales into the number of calves being slaughtered for dairy farmers by local hunts. This revealed that between July 1999 and the end of February 2000 hunt kennels received 14,688 bull calves, an average of 2448 calves a month. Both the Union and the magazine were concerned that the industry had lost a potential income of £2.2m from sales to veal producers in the rest of Europe at a time when producers were suffering major financial problems. Referring back to the concerns outlined above, however, it is also equally clear that these figures must conceal a great deal of heart-ache and moral questioning from producers who believe that the practice runs contrary to their sense of responsibility for their stock and to their standards as committed stockmen.

Economic necessity has dictated that a young animal be treated as a worthless commodity and that it be disposed of as waste. Experience has shown that the vast majority of people responsible for the care of livestock find the acceptance of this notion difficult to stomach and even harder to implement. One can only hope that the financial problems afflicting UK agriculture, and the livestock sector in particular, will eventually be resolved, and that producers who suffer these trials of conscience will again be able to apply the principle of respect to the managing of their stock.

In the meantime a number of measures could have been taken by the Government, which would have offered some degree of hope to a beleaguered industry. For example, in 'The Farmers Weekly' editorial of 20th. August 1999 it was suggested that the Government should consider 'reopening export markets, under carefully controlled conditions'. Whilst such a move would inevitably be met with a storm of protest from animal welfare groups, the price of £200 per calf, which was then being offered by veal producers, would have been a lifeline for many UK dairy farmers, as has already been suggested in the comment above. In responding to the animal welfare concerns, the editorial raised the question as to whether it was more welfare friendly to export the calves or to shoot them.

In addition I suggest that if the necessary welfare safeguards were in place, and if the export trade and foreign veal systems were effectively controlled and monitored, then calves for veal production could enjoy a life, albeit a short one, in which they would be treated as sentient creatures, and in which their essence would be fully respected.

'The Farmers Weekly' editorial also proposed a number of less controversial solutions to the problem. These included the suggestion that there should be a return to the use of dual-purpose breeds in dairy herds and the more widespread use of sexed semen on dairy cows. The first would enable the dairy farmer to produce bull calves suitable for fattening on for beef, whilst the latter would ensure that only heifers would be produced from Holstein-Friesian cows served by Artificial Insemination (AI). It was recognised, however, that neither of these solutions would provide the industry with the immediate help that the situation demanded.

Despite the imposition of measures as extreme as those outlined above, however, many believe that there has been too little too late and as a result question whether the UK livestock industry can or will survive. Indeed, it has been suggested that some sectors of the industry are already in a state of terminal decline ('Church Times' 3.3.00). It has also been forecast that the UK meat and dairy industries are in such a perilous condition that they will never again be able to compete on world markets against foreign competition.

As far back as 1997 the Foresight programme (1997:16), which was established by the government in partnership with the business and science communities, stated that 'no realistic scenario can, in our view, be based on the contention that British meat production and trading of home-produced products can prosper by attempting to compete in world markets on a cost and volume basis.' The authors' conclusion was that 'against a background of world markets and world production patterns, British meat production (no more than 2% of the total) should be regarded as a niche, made up of a number of smaller niches exploiting local production opportunities and directed at specific local domestic market demands.'

Despite the pessimism of their prophecy the report's authors assumed that livestock production would continue in the UK, even if the industry that remained was reduced and only supplying a niche, or quality market. Indeed, some would encourage such a development, believing that the high animal welfare standards, demanded of British producers, establish exactly the specialised trading advantage that the industry and the market requires (MAFF, 9.7.98).

3.5.2 Globalisation

In forecasting the reduced production and increased specialisation of the UK's livestock industry, the Foresight authors imply that British retailers will increasingly source their animal products from foreign suppliers. Indeed, the evidence since 1997, when the document was published, is that this is already happening in the pig and poultry sectors of the industry (MLC, 1999) and that it is likely to happen even more in the future. The increase in the import of pig meat products between 1998 and 1999 is particularly dramatic, as is illustrated in Table 3.

The threat of even more of the home meat market being lost to competition from cheaper imports was highlighted in an article in the 'Guardian Weekly' (Vidal, 9.3.2000:25) in which it was reported that 'global and European corporations, working on an unimaginable scale, are already targeting Britain.' Indeed, this has been a trading ambition of the American livestock industry for a number of years, with Hiron

(1996:18) reporting in the mid-1990s that their plans at that time were 'to raise their share of world trade in 2002 to be at 60.8% for poultry and 33% for pig meat.'

Table 3 Change in EU pigmeat imports (key categories) into the UK – 1998 to 1999 (adapted from NPA, 8.11.99:1)

Meat products	1998 imports '000 tonnes	1999 imports '000 tonnes	% increase
Frozen/chilled hams and shoulders	19	27	+43
Fresh/chilled meat (excl. carcases and hams)	23	29	+25
Frozen meat (excl.carcases and hams)	25	27	+12

This goal has major and disturbing implications, not only for those who are anxious about issues such as the food security of individual nations, but also for those who are concerned about animal welfare. Not only does a growing dependence on imported animal products make the monitoring of production and slaughtering systems more difficult, but also under World Trade Organisation (WTO) rules it makes the blocking of any imports on welfare grounds impossible. As Vidal (2000:25) noted, 'the reality is that four or five giant US-sized pig farms could produce most of the 1m tons of pig meat every year. Never mind the welfare standards or how the pigs are reared - under world trade rules which prevent protectionism, there is little that ministers can do except preside over the virtual end of an industry.'

Vidal's anxieties stem from the 1999 round of WTO talks, when one of the initial demands of the representatives from the European Union (EU) was for the inclusion of animal welfare issues in the debate on agricultural trade ('Farmers Weekly' 2.10.98:20). Unfortunately this demand became one of the early casualties of the discussions. As a result of the EU failing to secure any place for animal welfare concerns in the agreement, it is now illegal for any WTO member nation to block the importing of animal products from another member nation on welfare grounds even if those animals have been reared in systems deemed inhumane by the receiving country. I argue that this position is immoral and untenable.

If on ethical grounds the British consumer demands high standards of animal care from the nation's farmers, then consistency requires that they make the same demands on the producers of imported animal products, and that as their representatives the Government should implement this demand.

On these grounds I argue that the animal welfare issue should not have been so easily dismissed from the WTO talks and that the UK Government has an ongoing moral obligation to continue pursuing the matter both within the WTO and within Europe. Indeed, this is nothing less than was promised by the Junior Farm Minister, Elliott Morley, who is on record as stating that there is 'a case for banning some imports from countries using production methods regarded as cruel, or where animals were treated with growth hormones and chemicals banned in the UK' ('Farmers Weekly', 2.10.98:20).

As has been argued previously, there can be no moral justification for farmers having to endure such problems as have been imposed on the British pig industry. Forced by the Government to submit to strict and expensive welfare requirements they have had to watch their industry being destroyed by cheaply produced pig meat imports from systems that are illegal in this country. If animal welfare is a moral issue that demands national legislation, then morally that nation should stand by its principles and exclude products from any unacceptable production systems. To prohibit such a stance is not to advocate free trade but to permit moral licence.

3.5.3 Vegetarianism and the birth of the Animal Rights Movement

Livestock farming in the UK is predominantly concerned with rearing animals for people to eat as meat, eggs or dairy products. As a result it only remains commercially viable as long as the public continue to accept that animals can be used in this way. If this were to change then the livestock industry would collapse. The Animal Rights Movement and many vegetarians are committed to achieving that end.

Although the statistics for estimated meat consumption per capita in the United Kingdom show little change since 1987 (see Table 4) there is growing evidence to suggest that an increasing number of people are embracing vegetarianism. From the results of the 1993 Realeat Survey (Donnellan, 1994:7), however, it would seem that this increase is largely a phenomenon of certain socio-economic groups. The results show, for example, that between 1990 and 1993 there was a 40% growth in the number of vegetarians in the C2 socio-economic group and a 66% growth amongst those in the DE group.

**Table 4.   Estimated UK meat consumption:
1987 to 1996 (kg/head)** (Source: ADAS 1997:25)

Year	Beef	Mutton and veal	Pork and lamb	Bacon	Poultry	Total meat
1987	20.1	6.3	13.4	7.8	18.6	66.2
1988	19.0	6.5	14.0	7.8	21.5	68.8
1989	18.5	7.2	13.4	7.8	21.4	68.3
1990	17.3	7.5	13.5	7.5	22.3	68.1
1991	17.5	7.3	13.5	7.3	23.0	68.6
1992	17.2	6.5	13.4	6.9	24.5	68.5
1993	15.5	5.8	14.0	6.9	25.5	67.7
1994	15.8	5.9	13.8	7.1	26.7	69.3
1995	15.3	6.1	13.0	7.2	27.0	68.6
1996	12.5	6.4	13.7	7.2	27.9	67.7

This contrasts with an estimated national growth in vegetarianism over this period of 16% (Independence, 1994:26). From this it would seem that Adrian Harrison ('Food Matters World-wide', No.13) is probably correct in suggesting that 'vegetarianism as a virtue in its own right is largely a British, urban phenomenon ... in the developing countries which are, of course, still largely rural, the British-style vegetarian ethic is virtually unknown.'

Whilst vegetarianism may indeed be a modern urban phenomenon it is still important to recognise that a non-meat diet has been chosen by or forced on individuals and groups for many centuries. For a few, including a number of early Christian hermits, it was an ascetic choice made as part of a wider spiritual discipline. For others, such as the thirteenth century Cathari sect in France, it was part of their ethical and theological belief system (Sorrell, 1988:77).

For the majority, however, it was forced on them simply because they were unable to afford meat and as a result had to live on the cheaper vegetables, cereals and pulses that made up their subsistence diet. The combination of improved farming methods and the wider distribution of wealth over the last hundred years, however, has resulted in social changes which have brought about the almost universal acceptance and enjoyment of regular meat eating. At the same time, however, it has to be recognised that there has been growing concern about the methods used for producing that meat and, indeed, growing questioning about the reasons for even eating meat.

The actual number of people who are vegetarians is difficult to ascertain, as there are no statistics which indicate the proportion of the population which follows either an omnivorous or a vegetarian diet. 'The MLC (Meat and Livestock Commission), citing the National Food Survey, says that 97 per cent of British people are meat eaters and only 3 per cent are vegetarians, whilst the Vegetarian Society claims that 7.5 per cent (about 4.5 million) are at least semi-vegetarian (i.e. they eat no red meat) and 4.5 per cent have renounced all flesh, including fish' (Girling, 1997:32).

What is clear is that there has been a shift from the consumption of red meat in favour of white meat. Whilst it might be argued that this is an expression of welfare concerns it seems more likely that it is a result of poultry meat becoming cheaper over recent years rather than consumers being anxious over production systems (NFU, 1998:44). Red meat consumption has also been drastically affected by fears arising from the outbreak of Bovine Spongiform Encephalopathy in UK cattle in 1985. These fears were intensified by the Government's announcement in March 1996 that there was a possible causal link

81

between BSE in cattle and variant Creutzfeldt-Jakob Disease in humans. This resulted in international concern over the safety of British beef, its exclusion from foreign markets, and a dramatic reduction in home consumption (See Table 5).

Table 5. Estimated UK consumption of beef by usage:
1994 to 1997 ('000 tonnes product weight)
(Source: NFU, 1998:46)

	1994	1995	1996	1997
Household - fresh	570	534	447	492
Household - processed	235	246	189	241
Catering	250	256	186	217
Total	1055	1036	822	949

So dramatic was the public's immediate reaction that in the week following the Government's announcement livestock markets witnessed a 94% drop in the sale of cattle ('Farming News' 29.3.96). As a result it is possible to surmise that economic and health considerations have played a greater part in turning people away from eating red meat rather than ethical arguments in favour of vegetarianism.

Although it is clear from Table 5 that there is a regaining of confidence in British beef it is impossible to make any projection regarding future consumption. What is becoming obvious, however, is that there is a definite correlation in Britain between the age of the consumer and their meat eating habits (NFU, 1998:41). The correct interpretation of this trend could be critical for the future of UK livestock production. This is recognised by the National Farmers Union: 'If lower meat eating is associated with generation there is clearly an urgent need to ask why this may be happening. Only a cohort study that traced a panel of consumers through time would begin to give an answer to this important question' (NFU, 1998:41).

Whilst there is no evidence to suggest that such research has been initiated, there are statistics available which indicate that the young, and especially young women, are adopting vegetarianism. An article in 'The Financial Times' in January 1994 cites a report commissioned by ASDA in 1993 in which 11 per cent of a sample of girls, aged between 13 and 15 years declared themselves to be vegetarian. Similarly, a Gallup survey, which was also conducted in 1993, found that 13 per cent of a sample of females aged between 16 and 24 years were vegetarian ('The Financial Times', 1994). These results indicate a far higher preference for vegetarianism amongst young women than is found in the rest of the population (see above).

It may be that this is purely a feature of the current youth culture and social climate, with few young people having anything to do with farming and especially livestock. It may also reflect at best an over-developed sensitivity in the young or at worse an overt sentimentality towards animals, which is based on gross anthropomorphism. If, however, this trend is a deliberate rejection of meat eating based on the belief that it is wrong to kill animals then it poses major and possibly disastrous implications for the future of the livestock industry. A trend motivated by such beliefs must logically lead from a rejection of meat eating to the total rejection of all animal products. The reason for this is that those who could class themselves as semi-vegetarians, or lacto-ovo vegetarians (i.e. who eat dairy or poultry products) must ultimately find themselves in an untenable logical position which can only be resolved by accepting the vegan's point of view which rejects the eating of any animal product.

The links between the dairy industry and beef production, for example, are unavoidable as dairy cows produce calves, and the only outlet for unwanted bull calves is for the veal or beef trade. Dairy production is thus tied inextricably to meat production and must therefore be unacceptable to the committed vegetarian who is motivated by welfare concerns. In much the same way, that individual must not only question the use of certain methods used in egg production but must also question the production itself. Practical economics demand that older birds have to be culled. Efficient and profitable egg production requires the removal and therefore the killing of unwanted birds, an unacceptable position for those whose vegetarianism is based on the belief that it is wrong to kill any animal.

From the above it would seem, therefore, that veganism is the logical consequence of a vegetarian diet motivated by ethical concerns. This indeed was the conclusion that the writer of an article in 'The Times Magazine' came to: 'If you believe it is morally wrong to cause suffering to a farmed animal, then the moral must be absolute... If meat is murder, then so is a cheese sandwich' (Girling, 1996:32).

Others would contend, however, that the vegan's argument ignores the nature of the world in which we live. An article produced by the Meat and Livestock Commission draws attention to this by quoting from David Attenborough's justification for his wild life programmes. 'Humans are part of the whole ecosystem and will take a life in order to sustain their own. They have done so since humans came into being. They have the teeth of an omnivore and the gut certainly not of the vegetarian. People have become divorced from the realities of nature in their urban environment. I hope to bring back in my programmes a clear understanding that we are part of that wider system and that animals die and are eaten' (Donnellan, 1994:23). This acceptance of the world as it is, is reflected in a further quote in that same article from David Bellamy: I'm not a vegetarian because more than a third of all our agricultural land can only produce grass. If I could eat grass and digest it I would, but I cannot - and as an ecologist, I say we have to utilise animals' (Donnellan, 1994:23).

The ethical concerns that motivate the stance taken by vegetarians will be discussed more fully in Chapter 4 along with the issues raised by the philosophy of animal rights/liberation. Before that can be achieved however, the practical influence of the Animal Rights Movement and its role in promoting ethically motivated vegetarianism needs to be recognised. The link between the two movements is umbilical, in that one cannot exist without the other.

Much of the impetus that gave birth to current animal rights/liberation activism arose out of the general concern for the rights of exploited individuals and groups that was such a major feature of life in the 1960s. The philosophies that inspired the Black Rights Marchers or the Gay Liberationists led inexorably into the philosophies propounded by Singer and Regan. This succession is recognised by Singer in his preface to 'Animal Liberation' where he acknowledges that 'the liberation movements of the Sixties had made Animal Liberation an obvious next step' (Singer, 1995:xvi).

Whilst the 1960s may have given birth to a populist movement to promote animal rights, the under-girding philosophy had much earlier origins. Singer (1995:xvi) traces it back to Henry Salt's book 'Animal Rights' in 1892, although Salt was himself influenced by earlier writers including Primatt, Lawrence and Bentham. It was in fact in the seventeenth century that the existence of rights was first proposed with the emergence of the School of Natural Rights. Its exponents included Grotius, von Pupendorf and Thomasius, who held that if such rights did exist then they were exclusive to humanity.

It was to take another century before the possibility of extending the concept to animals was first postulated in the 'System of Moral Philosophy', published by the Scottish philosopher Francis Hutcheson in the early 1700s (Maehle, 1994:92). His views were limited to the proposition that animals have a right not to have any unnecessary pain inflicted on them, but they mark the beginning of an animal rights philosophy. Henry Primatt developed this further and declared in 1776 that animals share with us a right to happiness and that the criterion for establishing that right should be their ability to experience pain rather than their possessing rationality (Maehle, 1994:93). In this, he predates Jeremy Bentham, who in 1780 proposed the same criterion in his 'Introduction to the Principles of Morals and Legislation' (see Singer, 1995:7). In suggesting that animals possess certain rights, both writers provided an alternative to the philosophies of Alpinus, Hume and Kant who extolled the indirect duties that they believed humanity has to acknowledge in all dealings with animals.

A further example of this concern for the subject of animal rights can be found in Lawrence's 'Philosophical and Practical Treatise on Horses' which was published in the 1790s and which included a chapter on 'The Rights of Beasts' (Maehle, 1994:94). Likewise, Thomas Young's 'Essay on Humanity to Animals', which was published in 1798, followed a similar line (Maehle, 1994:97).

Such works set the tone for much that was to develop into the animal rights/liberation movement in the latter part of the twentieth century, although as Singer notes in writing of Salt's book 'Animal Rights': it was 'left to gather dust on the shelves of the British Museum library, until eighty years later, a new generation formulated the arguments afresh' (Singer, 1995:xvi). When this happened one practical implication that became clear was that vegetarianism and animal rights are inseparable. If it is accepted that animals have rights, then it becomes difficult to justify their use as a source for food. As will emerge later, Singer as a utilitarian rarely refers to animals possessing rights and prefers to use the concept 'equality of consideration'. As a result, whilst he argues forcefully for the adoption of the vegetarian principle, his views are not so emphatic as those of the animal rights school. Regan (1988:351) however, is in no doubt of the dietary consequences of his views: 'on the rights view, vegetarianism is morally obligatory... and we should not be satisfied with anything less than the total dissolution of commercial animal agriculture as we know it, whether modern factory farms or otherwise.'

This statement emphasises how difficult it is for welfarists and animal rights activists to work together, especially with regard to farming issues. At first sight it would seem that, as both appear to be committed to the alleviating of suffering amongst animals they must therefore share similar aims. Such an understanding, however, is illusory. Whilst the former seeks to improve the conditions under which farm animals are reared, the other hopes to ban the actual rearing process itself. Whilst one accepts the motives for producing the animals, the other is totally opposed to them. The reason for this wide divergence between the two views is very basic, in that each starts from a very different understanding of the human/animal relationship. One believes that the fundamental differences between humanity and other species are so profound that humanity is in a unique and separate category to all other forms of life. The other minimises those differences, plays down that uniqueness and consequently emphasises a commonality that transcends the species barrier.

The debate as to which of these two conflicting schools of thought is right will be considered in Chapter 4. The outcome of that debate should determine whether or not it is possible to establish an acceptable ethic for livestock production.

CHAPTER 4

Human/Animal Relationships

4.1 Sentimentality or sympathy?

Having identified the sense of 'being overwhelmed by change' as a malaise of contemporary life Toffler, (1973:11) argues that 'unless man quickly learns to control the rate of change in his personal affairs as well as society at large, we are doomed to a massive adaptational breakdown.' Indeed, the emergence of subcultures is an indication that this is already happening. The sense of 'loneliness, alienation and ineffectuality' which Toffler believes to be endemic in late twentieth century society is being resolved by individuals 'plugging in to one or more of their subcults' as a hope of maintaining 'some sense of identity and contact with the whole' (Toffler, 1973:282).

One effect of such 'subcult affiliation' was identified as a growing alienation between group members and those who remain outside of it. Indeed, Toffler (1973:283) suggested that those who refused to conform to the ideas or practices of a group or who disagreed with that group's values or beliefs were likely to be met with ridicule and caricature rather than well-argued intellectual debate. Motivated by the desire to uphold the identity and objectives of the group rather than open them to doubt and question the arguments and views of the opposition are summarily dismissed.

Elements of the confrontation between livestock producers and the advocates of animal rights over recent decades would appear to confirm something of Toffler's view, as members from both groups, which may easily be identified as subcults, have been known to use dismissive generalities and parody in their portrayal of the other side's case. The impression conveyed is that the individual's commitment to his own group has prevented giving any credence to the views of non-members. A farmer, for example, who blithely dismisses those opposed to current livestock production systems as 'cranky vegetarian minority pressure groups' ('Farmers Weekly' 11.9.98:91) appears to have failed, or possibly just refused, to recognise both the sincerity and the intellectual convictions that frequently give rise to such opposition. Likewise, whilst Budiansky (1994:128-130) appears to respect Singer's intellectual integrity, he is also ready to resort to generalisation and caricature when he describes the views of animal rights advocates as 'simplistic' and suggests that they are frequently motivated by a romanticised view of nature and a sentimental attitude towards other species. For some, this may indeed be true, but with regard to the serious advocate of animal rights or animal welfare, it is far from being the case. Indeed, in introducing 'Animal Liberation' Singer (1995:x) actually emphasises that his concerns are not motivated by either sentimentality about, or love for, animals. Describing how both he and his wife feel about other species, he writes, 'Neither of us had ever been inordinately fond of dogs, cats, or horses in the way that many people are. We didn't "love" animals. We simply wanted them treated as the independent sentient beings that they are, and not as a means to human ends.'

As will be discussed in Chapter 6, those who rear animals for human consumption have a responsibility not only towards their stock but also towards society in that the relationship so described implies at least the mutual interests and duties that exist between suppliers and their customers. Thus, if there are sound reasons for complaint against current practices and systems in the UK livestock industry, those who are engaged in that industry have a responsibility both to listen to the arguments and where necessary to engage in serious debate with the complainants rather than to dismiss or caricature them. Indeed, this much has already been appreciated by the NFU as the representative body for English farmers with their commitment 'to an open and full debate with consumers which ensures that reasonable and realistic aspirations on animal welfare are clearly defined and met, without compromising animal health' (NFU, March 2000:6).

This chapter seeks to further that debate by examining the various influences that have moulded current attitudes towards animals.

Fundamental to the study is the belief that the use of vague generalisations and ridicule are inappropriate ways to respond to the honest criticisms and concerns of welfarists and rights advocates alike. In his Preface to 'The Case for Animal Rights' Regan (1988:xii) asks his readers 'to test how well I have reasoned about these matters, even if - one might say, especially if - the conclusions reached are critical of what you do'. I believe that such an appeal to reason encourages both mutual respect and a greater understanding between opposing parties and that it consequently creates an environment in which reasoned and informed debate can take place. Conversely, the strident confrontationalism, which so often characterises the argument, results in a polarisation of views, which can only serve to drive the participants further apart.

This point is emphasised by Midgley (1986) in her paper 'Conflicts and Inconsistencies in Animal Welfare' when she appeals for an end to 'mindless tribal hostility' between those of different views and for the developing of a closer co-operation between the different sides in the debate. She goes so far as to suggest that 'we need to internalise this debate, to feel both parties in it within ourselves. Inside the moderate, there should always be an extremist struggling to get out, and inside the extremist, there should be an equally restless moderate' (Midgley, 1986:3).

I believe that with this suggestion Midgley has overcome the problem identified by Budiansky regarding the way in which opposition to livestock production methods can too easily be couched in sentimentality. Indeed, in encouraging mutual understanding and respect between opposing parties, Midgely is arguing for what I would describe as sympathy in the place of sentimentality. Whereas the latter is defined as 'exaggerated, mawkish emotion' (Collins, 1996) and as such is readily expressed in the situations of 'mindless tribal hostility', which Midgely has referred to, the definition of the former is 'compassion for someone's pain or distress' (Collins 1996). By definition therefore, sympathy is not only bound to encourage understanding between opposing views but when applied to the experiences of animals in livestock production systems is likely to stimulate greater awareness and understanding of their needs and condition.

Scruton (1996:55) takes this further and suggests that sympathy is one of the four key components of moral action, along with virtue, personality and piety, and that it should be an integral feature of our relationship with other species. He states that 'real sympathy obliges us to know animals for what they are, to regard their bad points as well as their good and to take an undeceived approach to their suffering.' In contrast, he dismisses sentimentality as a 'form of self-conscious play-acting... The unreal love of the sentimentalist reaches no further than the self, and gives precedence to pleasures and pains of its own, or else invents for itself a gratifying image of the pleasures and pains of its object' (Scruton, 1996:100). On this basis, he concludes that sentimentality should be excluded from any moral debate.

As has already been indicated, whilst some of the critics of current livestock farming practices may appeal to sentimentality, most, according to Scruton's definition of sympathy, are genuinely motivated by a sympathy and concern for animals. In a similar vein many livestock producers would argue that any responsible stockperson or livestock management system must place due emphasis on the importance of sympathy regarding the care of livestock. Indeed, some commentators would argue that sympathy encompasses a number of the essential attributes required by anyone desiring to work with animals. I would suggest for example that Seabrook (1996:3) is implying that sympathy is a major component of stockmanship when he states that 'whilst the animal's perception of the stockperson is crucial, there is a necessity for the stockperson to attempt to understand the "world" from the animal's viewpoint and to understand how the animal perceives humans.' It is only when this sympathetic understanding is achieved that the stockperson can establish procedures for approaching and handling stock without causing them undue stress or fear.

Similarly, in describing the personal qualities needed by a stockperson, English *et al.* (1992:15) write of how any individual entrusted with the care of stock must have 'the ability and willingness to communicate and develop a good relationship to the stock.' This sympathetic approach to the animal will enable understanding to develop between them and their handler. As will be argued in Chapter 5, any ethically acceptable livestock production system must involve the exercising of the highest achievable standards of stockmanship and the consequent forming of a good working relationship between stockperson and stock. If this is the case then it must follow that the presence of a sympathetic attitude in the stockperson concerned must be equally essential.

Whilst sentimentality might therefore be excluded from the discussion concerning the moral implications of the relationship between humans and other animals the debate would be ethically deficient if the quality of sympathy were to be excluded from it. Rather, I contend that sympathy should be a major feature of any morally acceptable position with regard to how we treat and relate to other species as well as to each other. As Scruton (1996:34) has commented, 'Two of our sympathetic feelings are of great moral importance: pity towards those who suffer, and pleasure in another's joy. And these two feelings lie at the root of our moral duties towards animals.'

4.2 Animal rights or human responsibility?

Whilst it is recognised that the beliefs and actions of the various groups and individuals concerned with the well-being of animals may be motivated by a common degree of sympathy, there is a marked difference between the principles that stimulate that motivation. That difference is summarised in the above question. The sharp distinction between the two points of view marks the demarcation between those who believe that it is morally acceptable for humans to use other species for their own purposes providing they do so in a responsible way and those who believe that any such usage is morally reprehensible. It thus becomes the distinction between those who support the cause of animal welfare and those who advocate the principle of animal rights.

4.2.1 Can animals possess rights?

As has already been discussed in the previous chapter, animal rights advocacy has a history that stretches back into the eighteenth century. This relatively recent appearance on the philosophical scene is hardly surprising as it was only during the seventeenth and eighteenth centuries that the concept of human rights became a moral and political force, finding expression in the English Bill of Rights (1689), the Bill of Rights of Virginia (1776) and the French Declaration of the Rights of Man and the Citizen (1789) (see Rivers, 1997:1). It seems likely that the first to apply the concept to non-human species was Hutcheson in his 'Systems of Moral Philosophy', which was published posthumously in 1755. In this work he declared that brutes 'have a right that no useless pain or misery should be inflicted on them' (cited by Thomas, 1984:179). Hutcheson's views were then further developed in the eighteenth century by Primatt in his 'The Duty of Mercy' (1776) and by Lawrence in his 'Philosophical and Practical Treatise on Horses' (1796-78).

The animal rights cause was carried forward in the nineteenth century by Henry Salt who in 1894 declared: 'It is only by the spread of the same democratic spirit that animals can enjoy the 'rights' for which even men have for so long struggled in vain. The emancipation of men from cruelty and injustice will bring with it in due course the emancipation of animals also. The two reforms are inseparably connected, and neither can be fully realised alone' (cited by Thomas, 1984:185).

It wasn't until the 1960s and '70s, however, that the idea that animals might possess rights became more widely accepted, especially in urbanised society. Some of the historical and cultural reasons for this surge in interest and popularity have already been covered in the previous chapter but the pivotal role of Singer's 'Animal Liberation', which was published in 1975, is particularly significant. It was this book which became the popular focus of much of the debate on animal issues and consequently the philosophical reference point for many individuals and groups fighting against those human activities which they believed to be exploitative of non-human species. Indeed, DeGrazia suggested that 'more than any other work, Peter Singer's 'Animal Liberation' brought questions about the moral status of animals into intellectual respectability' (DeGrazia, 1996:2), whilst Leahy credited it with galvanising 'contemporary enthusiasm in topics with a considerable history, but until then, a low profile' (Leahy, 1991:14).

It is ironic, however, that whilst Singer is often cited as a key figure in the animal rights movement (Webster, 1994:6) he actually saw no reason for introducing the concept of rights into the moral debate concerning the status of animals. Indeed, his concern was for other species to be granted 'equal consideration of interests' (Singer, 1995:22) rather than rights. Regan (1988:219) illustrated this by quoting from Singer's paper 'The Parable of the Fox and the Unliberated Animals': 'Why is it surprising that I have little to say about the nature of rights? It would only be surprising to one who assumes that my case for animal liberation is based upon rights and, in particular, upon the idea of extending rights to animals. But this is not my position at all. I have little to say about rights because rights are not important to my argument.' Instead, Singer (1995:7) follows Bentham's line of argument that 'the capacity for suffering' is the vital characteristic on which the principle of according equal consideration of interests is based. He develops this further by suggesting that 'the capacity for suffering and enjoyment is a prerequisite for having interests at all, a condition that must be satisfied before we can speak of interests in a meaningful way'. He then proceeds to establish grounds for accepting that many other species share this capacity with humanity and goes on to argue that their suffering and enjoyment should therefore be accorded equal consideration with ours. When such consideration is withheld, he argues that the individual concerned is guilty of speciesism. 'To discriminate against beings solely on account of their species is a form of prejudice, immoral and indefensible in the same way that that discrimination on the basis of race is immoral and indefensible' (Singer, 1995:243).

Although Singer's conclusions have been widely acclaimed, they have also been widely criticised on a number of counts. Some have argued for example that an animal's capacity to experience pain and pleasure is insufficient reason for establishing a principle of equal consideration of interests between sentient species. This criticism was raised by Webster (1994:6) who, whilst acknowledging his admiration for Singer and admitting to sharing many of Singer's concerns regarding the humane treatment of animals, has also argued that the vast differences between the human and animal perceptions of suffering make it impossible to establish any equality of interests.

Indeed, he declares that as a result of these differences 'we must avoid the anthropomorphic projection of our own conception of suffering onto other species' (Webster, 1994:21).

To argue for an equal consideration of interests on the grounds that both humans and animals can experience pain is to suggest that those experiences are very similar. I suggest, with Webster, that this is not the case and that such experiences are often in reality very dissimilar. Whilst, for example, a cow with a broken leg appears to experience a similar degree of physical pain to a human with the same problem, it would be difficult to argue that they are suffering in the same way. Once the disabled cow lies down and has food put in front of her she is likely to chew contentedly on the cud and show little indication that she is suffering. A human with a broken leg, however, even when drugs have relieved their pain, is still likely to display evidence of distress, anxiety and even physical agony. They will probably feel a degree of anxiety about the effect of their disability on such matters as their plans, or their home and work responsibilities. They may be afraid of further pain associated with anticipated surgery, or they may experience a dread of extended hospitalisation. There may be fear of lasting disability or of the problems associated with having to learn how to walk again. The cow and the human share a common physical condition, but I suggest that they do not share a common experience of suffering and that their suffering cannot be described in terms of equal interests. Whilst sharing Singer's abhorrence of animal cruelty I find it difficult therefore to accept his argument that the ability to experience suffering makes for an equality of interests between sentient species. Indeed, I argue that the unique human capacity to anticipate, imagine, and worry means that there is little to equate the human experience of suffering with that of any other species (Leahy, 1991:256).

Furthermore, I suggest that this evidence of dissimilarity between the human experience of suffering and that of other species is equally valid with regard to the way in which the two experience pleasure. Whilst it may be said that the cow enjoys its food, its environment or its young there is little evidence to suggest that these experiences have the emotional, aesthetic and even spiritual connotations that they may have for a human who is enjoying them. Indeed the human experience would seem to be at a very different level from that of the animal. If, as I believe, consideration of equal interests implies that those interests must be of equal standing and value then the granting of equal consideration to the experiences of humans and other species on the grounds that they both experience pain and pleasure would seem to be difficult.

Singer's arguments regarding the relationship between humans and other species have been further criticised on the grounds that as humanity is uniquely different to all other species any argument for the consideration of equal interests is impossible to sustain. There can be no equal interests simply because there is no equality. Scruton, (1996:8) for example, in responding to Singer's views, states that this 'single-minded emphasis on the features which humans share with other animals - notably on the capacity for suffering - causes them to overlook the distinction between moral beings (to whom the argument is addressed) and the rest of nature.' In his arguments regarding the uniqueness and moral superiority of humanity over all other species, Scruton is following a long and ancient tradition. In Judeao-Christian culture, for example man is not only deemed to be created 'a little less than God', but to have had 'all things placed under his feet, all sheep and oxen, all the beasts of the field, the birds of the air and the fish of the sea' (Psalm 8). The highest status in creation is given to Man. In the New Testament this receives further emphasis in passages such as Matthew 10:31 in which Jesus is recorded as saying that man is of more value 'than many sparrows'.

Centuries later Kant argued for a distinction between humanity and all other animal species on the ground that man's ability to 'have the idea "I" raises him infinitely above all other beings living on earth. By this he is a *person* ... that is, a being altogether different in rank and dignity from *things* such as irrational animals, which we can dispose of as we please' (cited by Rolston, 1988:62). In asserting the uniqueness of humanity Rolston (1988:63) cites the work of the palaeontologist G.G. Simpson who argued that 'Man is an entirely new kind of animal in ways altogether fundamental for the understanding of his nature. It is important to realise that man is an animal, but it is even more important to realise that the essence of his unique nature lies precisely in those characteristics that are not shared with any other animal. His place in nature and its supreme significance to man are not defined by his animality but by his humanity... Man *is* the highest animal.'

In exploring the features of this uniqueness of humanity, referred to by Simpson, other writers identify a variety of differences between humans and other species. Scruton examines a number of these and suggests, for example, that the gulf between the two is one of morality, with humanity being the only moral species on earth (Scruton, 1996:36). The 'moral being' is defined not in terms of 'the rule-governed person who plays the game of rights and duties, but a creature of extended sympathies, motivated by love, admiration, shame and a host of other social emotions' (Scruton, 1996:32). Like Kant he also argues that humans are separated from all other creatures on the basis of their ability to 'refer to themselves as "I".' Humans are self-conscious and he argues that this is 'a feature of our mental life which does not seem to be shared by the lower animals' (Scruton, 1996:23). Rolston (1988:68) follows a similar line when he suggests that only humans are self-aware. 'Human self-awareness becomes intense and deeply reflective, abetted by philosophical doctrines and religious experiences that are absent from the animal world.' He later states that 'humans are part of the world in biological and ecological senses, but they are the only part of the world that can orientate itself with respect to a theory of it... They have a distinct metaphysical status just because they alone can do metaphysics' (Rolston, 1988:70).

Along with Wittgenstein, Frey and Leahy, Scruton sees further marks of uniqueness in our ability to use language (Scruton, 1996:25), whilst arguing that our 'piety' further distances us from other species in that it appears to be a quality unknown in any creature other than man. Scruton defines piety in terms of 'the deep down recognition of our frailty and dependence, the acknowledgement that the burden we inherit cannot be sustained unaided, the disposition to give thanks for our existence and reverence to the world on which we depend, and the sense of the unfathomable mystery which surrounds our coming to be and our passing away' (Scruton, 1996:55).

Whilst Singer (1995:238/9) contests all such claims to human uniqueness and disputes the existence of any 'concrete difference between human beings and animals' I would still argue that there is sufficient disquiet regarding his principle of equal consideration of interests amongst sentient species to suggest that it is inadequate as a basis for the relationship between humans and other species. Indeed, I believe that the differences between the two are such that any notion of equality becomes difficult to define and sustain. In following the same line Rolston (1988:74) has argued that, as there is an inequality of interests between species, 'there cannot (therefore) be equal consideration of unequal interests.'

I would further suggest that the above arguments regarding the uniqueness of the human species also imply that humanity has a unique role regarding responsibility in and for the rest of creation. It has already been suggested that no other creature exhibits evidence of a moral conscience. A cat, which takes time in the killing of its prey and plays with a mouse or bird by slowly despatching it by degrees, is behaving as a cat; it is what cats do. A human who behaves in a similar way, however, is said to be inhumane, their cruelty is condemned and their behaviour described as perverted. The cat is amoral and has no understanding of what morality means. As a result it would be wrong for human society to have any moral expectations of it, indeed it would be ridiculous to have such expectations. That same society, however, would expect the human in this illustration to exhibit some degree of moral conscience and to show some element of remorse or regret for their cruel behaviour. Sympathy and compassion are deemed to be humane qualities, integral parts of being human.

If it is true that humanity is the only moral species it must follow that no species other than humanity can feel or exercise a sense of duty for its environment or for its fellow creatures. Indeed, Rolston (1988:73) suggests that only humans 'can "look over" or "look out for" all other orders of life' whilst Warnock (1998:69) suggests that 'it is precisely because humans are fundamentally different from other animals that they have an obligation not to treat them with cruelty. Other animals have no such obligation.' As will be recognised in Chapter 5, this implicit responsibility has long been a traditional as well as an integral feature of our relationship with animals, especially in the world of livestock husbandry, and continues to be expressed in and through the qualities allied to stockmanship.

As was earlier recognised, Singer's influential voice in the debate concerning the relationship between humans and other species has often been wrongly identified with that of the animal rights advocates. In contrast I suggest that the description of Regan as the founding father of the animal rights movement is in many ways correct. Indeed, Regan's 'The Case for Animal Rights' (1984) has been described as the seminal work on the subject (Bruce 1998:128) and has been said by DeGrazia (1996:5) to be 'perhaps the most systematic and explicitly worked-out book in animal ethics.'

Whilst Singer rejects any suggestion that animals possess rights *per se*, Regan focuses on the concept of 'inherent value' which leads inextricably to his assertion that other species not only possess rights but that humanity has a duty to respect them. Inherent value is identified as the value that individuals 'have in themselves', which Regan suggests is distinct from 'intrinsic value that attaches to the experiences they have' (Regan, 1988:235). He further argues that inherent value is possessed by all beings that are 'subjects-of-a-life'. These subjects are identified as individuals which 'have beliefs and desires; perception, memory, and a sense of the future, including their own future; an emotional life with feelings of pleasure and pain; preference and welfare-interests; the ability to initiate action in pursuit of their desires and goals; a physcophysical identity over time; and an individual welfare in the sense that their experiential life fares well or ill for them, logically independently of their utility for others and logically independently of their being the object of anyone else's interests. Those who satisfy the subject-of-a-life criterion themselves have a distinctive kind of value - inherent value' (Regan, 1988:243).

In this one statement, Regan establishes his basis for determining both moral worth and the moral community, for just as some creatures are clearly excluded from that community in that they do not possess this inherent value, so others are clearly to be included because they do. His contention from this is that we owe a duty of justice and respect towards all those that meet the criterion although it has to be said that it is not always clear as to what this duty entails. On those occasions when it is defined, it tends to be in terms of not harming other subjects of a life. So, for example, he argues that, 'We have... a *prima facie* direct duty not to harm those individuals who have an experiential welfare' (Regan, 1988:262). Indeed, DeGrazia (1996:5) summarised Regan's views in these terms when he stated that 'inherent value implies a basic Respect Principle, which in turn implies "subjects".' This in turn 'leads Regan to the thesis that "subjects" have a right not to be harmed.' This then becomes the starting point from which Regan reaches his conclusion that humans are not alone in possessing rights and that rights are in fact the prerogative of all sentient life.

Whilst DeGrazia admits to admiring the thoroughness of Regan's arguments and the strength of his convictions, he remains unconvinced by Regan's conclusions regarding animals' possessing rights. Many others have shared and expressed DeGrazia's doubts regarding the validity of Regan's arguments including Leahy (1991), Carruthers (1992) and Scruton (1996). Some of those arguments and the criticisms directed against them will be examined in section 4.2.2, but I suggest that the notion of rights *per se* first has to be examined. One reason for this is that the use of rights language has become so popularised that it is easy to make assumptions about the definition, possession, scale and validity of rights. Rivers (1997:1), for example, has suggested that 'The language of rights has become endemic within Western societies' and that it 'has become the *lingua franca* of modern moral discourse.' He later suggested that 'The proliferation of rights-language has become increasingly bizarre and hence useless in the resolution of moral disagreement' (Rivers, 1997:4).

Whilst Rivers failed to provide any evidence in support of his statements, his concerns regarding the use of rights language echo the anxieties of other writers. Midgley, for example, declared in an interview for 'The Times Higher' that 'It is rather depressing how the term "rights" has been made an idol. There are serious objections to it which have been made over and over again' (Irwin 1996:15). Or, again, Warnock (1998:56), having defined a right as 'something you can claim, and which you can properly prevent other people from infringing... an area of freedom for an individual which someone else must allow him to exercise as a matter of justice', then proceeds to draw attention to the confusion that easily arises from 'the inextricable weaving together of legal and moral terminology in the discussion of the human rights issue' (Warnock, 1998:64). Her conclusion is that 'if the rights are moral, not legal, and if the law conferring them is a moral law, then the necessity for using the language of rights seems to disappear. Why should we not prefer simply to talk about ways in which it would be right or wrong, good or bad, to treat our fellow humans? This would be to adopt the language of morality itself, with no quasi-legal implications' (Warnock, 1998:63). I find this conclusion eminently sensible and practical, especially as I share many of Warnock's reservations regarding the use of rights language as well as certain features of the concept itself.

Warnock quotes, for example, Bentham's conclusion that rights not conferred by law are 'nonsense on stilts' and not rights at all. Her response to this statement is that 'there is in my opinion, a great deal to be said for this view' (Warnock, 1998:58). Bentham's view and Warnock's response to it indicate that they would find notions of inherent or intrinsic rights difficult if not impossible to sustain, in that such rights own no origin other than the existence of the one who possesses them. Their criticism of this notion of unconferred rights seems to me perfectly sensible. Rights require an origin, someone or something (e.g. a nation, community or organisation) to confer them, and I would thus agree with Tudge's statement (1997:40) that 'none of us are born with rights. They are not a property of flesh. We need not suppose that they are 'God given' and it is difficult to argue that they are 'inalienable'. Rights are a convenience, a social device, that makes it easier for us to live with each other.'

Some might argue that a belief in God overcomes Tudge's complaint, in that such belief allows for a divine conferring of rights on all or part of the created order. Whilst theologically this assumption may appear to be correct it has to be said that there is little evidence for this view within the Judaeo-Christian tradition. In fact there is no explicit biblical evidence to support the existence of inherent or intrinsic rights and any implicit references to any sort of rights at all appears to be at a purely legal level regarding property (Deut. 21:17, Jer. 32:7, Jer.32:8) or behaviour (e.g. I Cor. 6-9). Rolston (1988:47) actually suggests that not only is the idea of rights virtually unknown in the Bible, but that it is a largely secular concept. Indeed, one might add that the traditional emphases in the Christian faith have been divine grace and human responsibility and not rights. Rivers (1997:3) goes so far as to suggest that 'much modern rights-talk has connotations that are egoistic, licentious and antagonistic, in short that are profoundly anti-Christian. For rights carry an inherent bias favouring individualism over collectivism, autonomy over heteronomy, and conflict over consensus.'

Rivers' emphasis (1997:3) is on the flourishing of human relationships, which he suggests only happens 'when people not only give others their due, but spontaneously give them more than their due in countless acts of unrequested good will.' Warnock (1998) follows a similar line in suggesting that rather than talk about a matter such as children's rights, society should focus on humane and responsible adult behaviour. 'It is my contention that we should not think of such moral convictions as these, which lie behind the law and without which the law would not be changed, in terms of rights. We can express them simply in terms of humanity, the way we ought to treat human children' (Warnock, 1998:61).

In line with much of Warnock's reasoning I have found myself becoming increasingly dissatisfied with the language of rights and more convinced that the concepts and subsequent application of responsibility and respect are more conducive to the well-being and social and moral development of community. I would suggest that it is when people acknowledge and exercise a sense of responsibility for each other, instead of trying to assess the extent to which various individuals and groups possess rights, and what the nature of those rights might be that there is likely to be less aggression and confrontation within society and more compassion and care. I recognise that in certain instances this sense of responsibility might degenerate into interference or become patronising, but I suggest that the showing of respect would prevent this from happening.

Respect is defined in the Oxford Dictionary (1959) as 'to regard with deference; avoid degrading or insulting or injuring or interfering with or interrupting, treat with consideration.' When one individual respects another they are likely to show some degree of sensitivity towards the other's feelings, needs and aspirations and will not impose their will upon them. Indeed, I believe that this combination of responsibility and respect, which Christian theology summarises with the Greek word *agape*, is the fuel that enables relationships in community to function, whether those relationships exist between humans or between humans and other species. When either or both parts of the combination are absent then those relationships are likely to disintegrate.

This emphasis on responsibility and respect as the basis of inter-human and human/animal relationships, rather than the existence of rights, is found in the writings of a number of environmental ethicisits. Arguing that the concept of rights belongs purely in human society, they suggest that it is not only difficult to apply to the natural world but is completely absent from it and alien to it. Rolston (1988:48), for example, suggests that 'If we take persons off the scene entirely, in the wilderness the mountain lion is not violating the rights of the deer he slays. Animal rights are not natural in the sense that they exist in spontaneous nature. Rights go with legitimate claims and entitlements but there are no titles and laws that can be transgressed in the wilderness.' Having questioned the existence of inherent or intrinsic rights in this way he then proceeds to argue that the issue at stake is not about "rights" but 'about what is "right". The issues soon revert to what they always were, issues of right behaviour by moral agents' (Rolston, 1988:51). That right behaviour is then interpreted in terms of 'respectful appreciation' (Rolston, 1988:75) and responsibility.

Like Rivers, Page (1996:136), another environmental ethicist, is more concerned about 'quality of relationships' than rights and argues that this quality is 'more important than a list of rights to life and its conditions.' Her reason for rejecting the concept of rights in this way is that she believes that it lacks 'warmth' and that as a basis for relationships it is both unsatisfactory and unsatisfying. Along with Rolston, Page emphasises responsibility rather than rights as the ethical basis for humanity's relationship with the natural world and like him she expresses concern at the impracticality of conferring rights on all life forms. Page (1996:136) suggests, for example, that 'extending the notion of rights all the way to micro-organisms may trivialize the notion; the assertion of all these rights could lead to absurdities in practice; determining and balancing rights will be endlessly complicated; respect for the rights of non-humans is hopelessly impractical.'

In support of Regan's criteria for the possession of rights by non-human species I suggest that this accusation is unfair in that he assigns rights purely to sentient species, and within this to those creatures, which he defines as 'subjects-of-a-life' (Regan, 1988:243). Rolston argues, however, that even this distinction is impractical. There is some indication for example that earthworms experience a degree of sentience. Whilst they do not fulfil all of the criteria demanded for Regan's 'subjects of a life' they may well fulfil a number of them or even fulfil all of them in ways which we at present do not understand. So Rolston (1988:62) suggests that 'their form of sentience may be so alien to ours that "Do to others as you would have them do to you" is untranslatable. One hardly knows what universal benevolence, respecting their rights, would mean.'

These concerns echo some of the criticisms directed against the concept of animal rights by other writers. Warnock (1998:68), for example, argues that 'if all the animals had a right to freedom to live their lives without molestation, then someone would have to protect them from one another. But this is absurd.' Again in response it might be suggested that animals only possess rights in relation to humans, in that humanity alone has the ability to recognise and respect rights, and that as a consequence the question of protecting them from one another does not arise. For an environmentalist, however, this immediately raises problems in that human existence impinges on the life of most other species, and as a result the debate concerning animal rights must have implications for our dealings with most other mammalian life forms. There are times, for example, when human intrusion could save animals in the wild from a great deal of suffering. In situations of extreme weather conditions when food supplies are restricted by drought or flood, what are the animals' rights in relation to humans? Should Park Rangers or Wildlife Wardens feed them in the belief that they have a right to food and therefore survival, or should the right of those animals to live or die in their natural environment free of human interference be respected? The evolution of species and the reality of the survival of the fittest suggest one answer whilst the relief of suffering implies another.

The fact that human existence does influence the well-being and even survival of other species has further implications for the practical application of an animal rights argument. For example, it is impossible for even the most committed vegan or animal rights activist to live their lives without asserting their own survival over that of other creatures and without thus denying those creatures the rights which they are deemed to possess, or making those rights conditional. As many of those species, such as germs and infectious insects, do not comply with Regan's 'subjects of a life' criteria their necessary destruction cannot be considered an infringement of any possessed rights. There are many others, however, that could be classed as 'subjects-of-a-life' and whose subsequent rights would be infringed or over-ruled by human activities. As Midgley (1984:25) maintained 'Crop pests of all kinds – not just insects, but rodents, birds, even deer, baboons and elephants – must be killed, if only by starvation, by people who mean to survive, and this would be true even in a world of Jains.'

Webster (1994:9) made a similar point when in quoting from Burns' poem 'To a Mouse', which he describes as the 'first and best of all animal welfare poems', he refers to the destruction being wrought upon the mouse's habitat by the turning of the plough. The reality is that humanity's need for food, as well as housing, transportation and industry, will always threaten the existence of other creatures. Arable farming can be as destructive of animal life and as exploitative of any perceived animal rights as livestock production.

On the basis of these concerns regarding both the concept of rights and the impracticality of applying that concept to non-human species I find that it is an unsatisfactory ethical base on which to build the relationship between humans and animals.

4.2.2   Respect - a responsibility to care

If as I have suggested, the concept of animal rights is deficient as a moral basis for the relationship between humans and other animal species, then an alternative has to be found. As was illustrated in the previous chapter, the traditional western and thus Christian-based perception of this moral foundation stems from the divine command in the opening chapter of the book of Genesis for mankind to exercise 'dominion' over the rest of creation. Biblical passages such as Psalm 8 and Romans 1:20, define this in terms of an implicit respect for creation, a 'respect for all species in their proper place in the world' (Broom, 1989:95). This respect, as already claimed, is a direct consequence of man's appreciating that the created world is God's handiwork, and that each part intrinsically displays something of his character and purpose. Whilst the theological contribution to the debate *vis-à-vis* animals and humans will be addressed more fully in Section 4.3 and the notion of respect will be discussed more fully in Section 5.1.1, it has to be recognised at this stage that an implicit respect for other species not only provides an alternative moral basis for that relationship but that it also has secular as well as theological support. Just as much of the theological debate centres on a belief in the uniqueness of humanity which is described in the book of Genesis as our being 'made in the image of God', so too the secular argument as described below tends to focus on basic mental, emotional and spiritual differences between humans and the rest of the animal kingdom - differences that carry with them unique responsibilities for the well-being of the rest of the natural world.

Whereas some might fear that this view of humanity's status is likely to drive the debate into further exploitation and abuse of animals, which is then justified on the grounds of human superiority, this need not be the case. The natural corollary to an appreciation of human uniqueness is the recognition of the distinctiveness or 'specialness' of every other species. Each creature should be valued for what it is and for the unique nature and characteristics that it possesses. So whilst Rolston (1988) and Page (1996) both emphasise the uniqueness of humanity, as ecologists they also emphasise the distinctiveness of every other member and feature of the natural environment. Hence Rolston (1988:68) writes of each natural kind having 'its place integrity, even perfections' and suggests that 'a discriminating ethicist will insist on preserving the differing richness of valuational complexity, wherever it is found' (Rolston, 1988:66). Similarly Page (1996:xiii) writes of the 'web of creation' in which everything, whilst being unique, is inter-related, interdependent and thus to be valued.

As has already been recognised, whilst Singer and Regan both acknowledge the existence of some differences between humans and other members of the animal kingdom, they tend to blur the distinctions by emphasising equality between the species, be it equality of interests or equality of rights. In contrast, Scruton (1996), and Leahy (1994), both recognise the significance of those human qualities and characteristics that separate us from all other creatures. In their arguments, the demarcation line is very clear and there is no blurring of the cross-species barrier.

Leahy (1994:254), for example, suggests that the gulf between humans and the rest of the animal kingdom is vast. He writes, 'animals are primitive beings, far removed from ourselves, despite some apparent behaviour to the contrary.' He identifies three major areas of difference, 'We can talk, are moral agents and are self-conscious'. It is our ability to talk, however, that he focuses on as the main feature of human uniqueness. 'Animals, unlike plants, are conscious (they see, hear, feel and make noises) but self-consciousness like hope, ambition, remorse and true memory comes only with the capability of speech' (Leahy, 1994:255). Animals may be able to communicate in very primitive ways, but 'animal "signals", however interesting as a subject of study, are sharply different from speech.' Their communication is purely instinctive, 'merely a response without any thought process regarding the meaning of the message' (Leahy, 1994:33). So great, therefore, is the gulf between us and them, that it makes a nonsense of the claims to equality made by the advocates of animal rights. 'My contention is that our understanding of what animals really are has not been portrayed with rightful subtlety. In particular, the role of language in the equation has been grossly underestimated. The implications of its absence in animals permeate to the very heart of our everyday talk about them. The liberationists exemplify the varied confusions that abound when these are ignored' (Leahy, 1994:163).

The suggestion is that the liberationists' emphasis on species equality has actually resulted in a misunderstanding of the true nature of animals and thus a diminishing of the respect that they are due. Accordingly, he argues for a return to an Aristotelian understanding of humanity's social pre-eminence within a hierarchy of living beings, which in itself entails an acknowledgement of respect and responsibility towards other creatures. His conclusion is that 'an understanding of the true workings of animal life, particularly along the lines originally proposed by Aristotle and Aquinas, which will be seen to anticipate a great deal of contemporary ethology, can provide a firm foundation for treating animals humanely but at the same time eating them, experimenting upon them and a lot beside' (Leahy, 1994:76). Conversely, 'attempts to convince us that the eating of meat and fish is an evil invasion of the inalienable rights of animals and that it should cease forthwith are a sham. They can only succeed with the help either of opportunistic flights of fancy such as inherent value or *theos* rights, or by obscuring the differences between creatures like ourselves that use language, and those that do not. The result of so doing is the sad and mischievous error of seeing little or no moral difference between the painless killing of chickens and that of unwanted children' (Leahy, 1994:220).

Whilst Scruton (1996) shares the objections to animals' rights that are proposed by Leahy, his is a more qualified view as regards our responsibility towards other species, in that he applies his concerns to specific systems of livestock production. He notes, for example, how 'most people find the sight of pigs or chickens, reared under artificial light in tiny cages, in conditions more appropriate to vegetables than to animals, deeply disturbing, and this feeling ought surely to be respected as stemming from the primary source of moral emotion' (Scruton, 1996:82). In a similar vein, he is keen to emphasise that his arguments against the philosophy of animal rights are not to be interpreted as an excuse for human cruelty. 'All thinking people now recognise the gulf that exists between sentient and non-sentient beings and almost all recognise that we have no God-given right to ignore the suffering that we cause, just because the victim belongs to some other species' (Scruton, 1996:7).

Just as Scruton recognises the existence of a gulf between sentient and non-sentient life forms, so too he identifies a marked distinction between humans and all other creatures. The quality he discerns as being distinctively human is the possession of a moral character. This immediately raises the issue not only of our moral pre-eminence, but also of the moral responsibility and duty towards other species which result from it. He writes, 'it is an empirical question whether we humans are the only moral beings on earth. I am inclined to believe that we are indeed alone in this respect' (Scruton, 1996:36).

As has already been stated, he suggests that our moral character consists of four components, namely 'personality, with its associated moral law; the ethic of virtue; sympathy and finally piety' (Scruton, 1996:55). Each is underpinned by reason and leads inevitably to an acceptance of duties and responsibilities towards other individuals and towards the wider community. He argues, for example, that the moral law imposes obligations or duties on all rational beings and that these duties include a responsibility towards non-human species (Scruton, 1996:53). The ethics of virtue 'ensure that people overcome the temptations posed by greed, self-interest and fear', and this 'makes a substantial contribution to the question of how we should treat animals. It compels us to distinguish virtuous and vicious attitudes towards other creatures regardless of whether those creatures are moral beings like us' (Scruton, 1996:54).

The role of sympathy has already been discussed, whilst that of piety, which he describes as 'the punctilious respect towards sacred things', he believes to be an essential part of the moral consciousness. 'It is piety, and not reason, that implants in us the respect for the world, for its past and its future, and which impedes us from pillaging all we can before the light of consciousness fails in us' (Scruton, 1996:57).

This moral consciousness, which is at the heart of our uniqueness, impels us therefore to treat other species with respect for what they are, and as a result to adopt a responsible attitude towards them. Consequently, if one accepts the objections to the case for animal rights, Scruton's perception of the relationship between humans and other animals would seem to be both a viable and practical moral alternative. Whilst the same may equally be said of the arguments put forward by Leahy, I would suggest that Scruton's approach to the relationship is particularly relevant to the world of livestock production. His emphasis, for example, on our responsibility to understand and respect animals for what they are as animals (Scruton, 1996:99) is very much in line with what has already been described as essential stockmanship.

108

Furthermore, Scruton's four-fold definition of the moral character as previously summarised provides some guidance as to how we should use and treat farm animals. His identification of sympathy as a key component in the moral character is in line with the common view that it is an essential feature of the quality of stockmanship.

4.3   The theological perspective

Although the era has long since passed when theology was known as the 'queen of the sciences', it is difficult if not impossible to engage in the debate regarding the relationship between humans and other species without recognising the influence that it has had, for good or ill, on that relationship. As has already been illustrated many of our perceptions in western society with regard to the status of other species are rooted in Christian culture in general and in Christian theology in particular. These influences have often been detrimental for animals in that some of the Church's most significant scholars, such as Aquinas, taught that non-humans are without souls and are therefore devoid of any moral significance (Singer, 1995:191). As Midgley and others have indicated, however, such an attitude has chosen to ignore the many biblical texts which emphasise human responsibility towards other creatures (Midgley, 1983).

Those who observed the scriptural injunctions to care and take responsibility for all of creation played an influential part in the birth and development of the animal welfare movement. Indeed, the very history of that movement in the UK illustrates a response to these scriptures, for the commitment of key figures such as Primatt and Broome to the welfare cause was directly attributable to their Christian convictions (Linzey 1994:17,19). This fact is appreciated by Linzey and Regan who recognised that 'religious involvement in animal-protection efforts was a matter of course in the eighteenth and nineteenth centuries' (Linzey and Regan, 1990:118-120).

Furthermore, whilst the influence of specifically Christian thinking on the wider life of society may have diminished during the second half of the twentieth century, religious and therefore theological concerns are still of general interest. It can be argued that this is especially true with regard to beliefs concerning the human/animal relationship. For example, the influence of a Christian writer such as Linzey on the animal rights debate cannot be understated, whilst the post 1960s interest in Eastern religions has undoubtedly had a profound effect on the way we relate to and treat other species. The apparent growth in the number of vegetarians since that era, discussed in Section 4.5, may be attributed to the growing popularity and influence of Buddhist and Hindu theologies in the west over that period of time.

As western society has become more pluralistic religiously, it has had to recognise the emergence of a number of different moral issues regarding the treatment of other species that arise out of the theology of a non-Christian faith such as Islam, which views and treats animals in ways that some might find abhorrent or distasteful. The establishing of strong Moslem communities within western societies, for example, has made an issue such as the ritual slaughter of animals a very relevant item on the agenda of animal welfare concerns within those host countries.

As regards the specifically Christian contribution to the debate on animal issues, however, no contemporary theologian has had a greater influence than Andrew Linzey, who has been described as 'the leading Christian theologian expounding a radical shift in the traditional view of animals' ('AgScene' Summer/Autumn 1994:23). Inspired by the work of Albert Schweitzer, and especially by Schweitzer's reverence for all living things (Linzey, 1994:4-7), Linzey argues that such reverence is 'the sole principle' of morality from which all other principles evolve (Linzey, 1994:4). Moreover, he maintains that this principle is a fundamental feature of the way of life taught and epitomised by Christ, and consequently he argues that human dominion over animals needs to take as its model the Christ-given paradigm of lordship manifest in service (Linzey, 1994:2-3). This is an essential ingredient in his theology and has led to the introduction of two major concepts into the debate, those of 'theos-rights' and the 'generosity paradigm'.

The idea of 'theos-rights' is based on the belief that all rights are dependent on the graciousness of God. 'Since God's nature is love, and since God loves creation, it follows that what is genuinely given and purposed by that love must acquire some right in relation to the Creator... The notion of 'theos-rights', then, for animals means that God rejoices in the lives of those differentiated beings in creation enlived by the Spirit. In short: If God is for them, we cannot be against them' (Linzey, 1994:25). Dissatisfaction with this definition is expressed by Leahy, who notes that it appears to emanate 'from the God of Christianity being on the side of or "for" his creation'. This is a concept which he dismisses as 'an insecure notion', arguing that there is little biblical evidence to support it (Leahy, 1994:211). I would add that, as the concept of rights is practically unknown in Christian scriptures, Linzey's introduction of the notion of 'theos-rights' is artificial and unnecessary. Indeed, his declaration that 'If God is for them, we cannot be against them' is perfectly acceptable without any appeal to rights. If God cares for and values his creation, then religious faith demands that the believer does the same, whether they acknowledge the existence and possession of rights or not. The believer's caring response is motivated by their religious faith and not by an acknowledgement that another species possesses divinely endowed rights.

Linzey draws his inspiration for the 'generosity paradigm' directly from his faith in Christ, who in exhibiting 'an inclusive understanding of divine generosity' sets the pattern for all personal relationships - be they with other creatures or with other people. The love of God which is expressed in Christ and which is available to all without limit or favour is profoundly generous. Such Christ-like love consequently becomes the hallmark of the Christian community (John 13: 34-35), and the basis of a self-sacrificial personal life style. 'I suggest that we are to be present to creation as Christ is present to us. When we speak of human superiority we speak of such a thing properly, only in so far as we speak not only of Christ-like lordship but also of Christ-like service. There can be no lordship without service and no service without lordship. Our special value in creation consists in our being of special value to others' (Linzey, 1994:33).

Whether, like Scruton and others, one views our relationship with the rest of creation in terms of responsibility, or like Linzey in terms of a Generosity Paradigm, the outcome would seem to be much the same.

The needs and the nature of other species are recognised, respected, and cared for. Indeed most Christians would consider Linzey's concept of the 'generosity paradigm' an integral part of their faith, as a natural expression of Christ's call to love and service. In this area of the debate, few Christians could dispute his argument, especially when he writes 'Drawing upon the notion of divine generosity exemplified in the person of Jesus, I suggest that the weak and defenceless should be given not equal but greater consideration. The weak should have moral priority' (Linzey, 1996:181). The problems arise for many Christians, however, when Linzey assigns this moral priority to members of non-human species. In his teaching on the pre-eminence of love, typified in the parable of the Good Samaritan, Christ is clearly dealing with our moral and spiritual responsibility towards other people, and makes no reference to any moral prioritisation of other creatures (Luke 10:30-37). In fact, in passages such as Matthew 6:26, he actually appears to do the opposite when he states that God values humans above the rest of his creation.

Whilst rejecting the notion of animal rights on the grounds discussed in the previous section, and finding it difficult to accept a belief in 'theos-rights' as anything more than 'a theological importation' into the debate (Leahy, 1996:189), I still find Linzey's work appealing and inspiring. He grapples with the paradoxical nature of the biblical teaching with regard to the status of animals in a way that few other theologians have dared, and his arguments and conclusions are always examined in the light of the biblical texts. Believing that 'the promise of real theology has always been that it will liberate us from humanocentrism, that is from a purely human view of the world to a truly God-centred one' (Linzey and Yamomoto, 1998:xvii), he endeavours to achieve that end.

Linzey exhibits a great sensitivity with regard to God's purposes for his creation, humanity's place within it and especially the suffering of animals that is so much a part of it. He emphasises how human responsibility is divinely ordained, and how it must be exercised according to the pattern established by Christ. Above all, he has achieved more than any other theologian in terms of making the Church aware of the moral and theological issues implicit in our relationship with other species and of our subsequent responsibility towards them under God. No other theologian of his calibre has yet emerged to present an alternative theological perspective and to answer his arguments.

Whilst Linzey has provided a theological advocacy for the animal rights cause, and has brought such issues to the attention of the Church, the ecclesiastical establishment has largely continued to follow traditional doctrinal views regarding the relationship between humans and the rest of the animal kingdom. For example, belief in the uniqueness of humanity as depicted in the creation story of Genesis chapter 1, where man is made in the image of God, remains a pivotal feature of Christian doctrine. 'Traditionally Christians accorded a unique place to humanity... being made in the image of God. Alone of all earthly beings, we possess reason and a capacity for self-transcendence, and thus for worship and prayer' (Methodist Conference Agenda, 1997:12). Likewise, belief in divinely ordained human dominion over the rest of the created order, as described in Genesis 1:26-28, is upheld and defended. Contrary to many popular views, however, this belief in human dominion is not seen as a licence for exploitation, but as an expression of divine trust in human responsibility and stewardship. 'It is integral to our Christian faith that this world is God's world and that man is a trustee and steward of God's creation who must render up an account for his stewardship. He must therefore exercise his dominion in conformity with God's will and purposes, not only in relation to himself but to the whole area of created life' (Archbishop of Canterbury, 1981).

Sadly, this dual Christian emphasis on the uniqueness and dominion of humanity has frequently been misunderstood by its opponents as excusing all manner of abuses and exploitation against the rest of creation. Porritt, of 'Friends of the Earth', for example, has stated that 'we shall continue to see a worsening ecological crisis until we reject that central Christian axiom that nature has no reason but to serve man' (ACORA, 1990:12). In the past, this may well have been the case, although it has to be said that Christian theology is far from being the only guilty party in seeking to justify such an attitude, as secular philosophies were also used to the same end. The fact is that over recent decades there has been a subtle shift in theological thinking.

Whilst humanity's dominion over creation is still acknowledged, and whilst our human uniqueness remains a fundamental feature of Christian theology, it would be wrong to assume that Porritt is correct in suggesting that the Church therefore believes that creation exists solely for human purposes and pleasures. The Archbishop's statement clearly rejects such a view, as does the Anglican report 'Faith in the Countryside' (ACORA, 1990). In this it is suggested that current Christian theology effectively reverses 'the normal western perception that the world exists for the sake of the human race; indeed - and paradoxically, because the human species is the most powerful - humanity is seen as existing for the sake of nature. 'Dominion' can be seen as the working out of the meaning of 'image'; an expression not of the exploitative powers of the human species, but of the loving power of God for the integrity of creation' (ACORA, 1990:13). Humanity, therefore, should be a reflection of God's love for his creation, and seen as his way of caring for it.

Furthermore, this belief in the 'loving power of God for the integrity of creation' must inevitably lead to a deep respect and reverence for all that God has created. As the believer worships God as Creator and reveres his purposes for that creation, so too he looks for evidence of the divine presence within it (Acts 14:15-18; Rom.1:20). Consequently, that believer's attitude towards the rest of the created world reflects a sense of worship, of awe, reverence and respect. Indeed, those elements of Celtic theology and spirituality that have had a profound influence on Christianity over recent years express such worship by embracing every area of life, including the welfare and husbandry of animals. There are prayers for shepherds and their sheep, for hunters and fishermen, for milking and lambing. All are characterised by this sense of God's presence in his world and its creatures, and yet they remain free of any maudlin sentimentality.

This is made clear by Adam (1985:60), who in describing the writing and collecting of such prayers, states 'We firmly believed in the Incarnation, so we sought a new earthiness in our prayers. They were to be truly grounded, they related to the lives we lived'. This earthiness can be defined as a respect for things as they are, made by God and devoid of human sentimental or emotional accretions and, as Muddiman (1998:30) indicates, it is found repeatedly in the teaching of Christ. 'There are no fables, everything is normal - sparrows drop dead, dogs scavenge and lick the wounds of beggars, and eagles gather over a carcass. Wolves can be expected to harass sheep - it would only be reprehensible if, like false prophets, they dressed up like sheep to do it. There is certainly realism here, and a respect for the way things are, since that is the way God has made them'.

In the previous section, the value of Scruton's understanding of the moral character and its various components for the establishing of an acceptable ethic for intensive livestock production was recognised. It is now possible to suggest that certain aspects of that understanding are also in accord with the theology outlined above. Scruton argues that the moral character consists of four components: i) personality with the associated moral law, ii) virtue, iii) sympathy and iv) respect or piety. Although there is a certain lack of clarity regarding various aspects of these components, for example there appears to be confusion between virtue and the 'virtues', especially as sympathy would be classed as one of the latter, there are still common points with the proposed theological view *vis-à-vis* the human/animal relationship. There is, for example, a respect for creation as it is, which naturally leads the believer to offer worship to the Creator even when confronted by the more unpleasant features of the created world. Similarly, the love for all of creation, and the readiness to care for it that is implied in the statement from the 'Faith in the Countryside' Report echoes the sympathy and related virtuous concern that Scruton identifies as belonging to the moral character.

All of this is ably summarised by Russell (1983), who, in the following statement, defines what I would hold to be the theological criteria for establishing the ethical acceptability of livestock production methods. 'The affirmation that the natural world is created by God leads to an appreciation of the essential goodness, order and holiness of all created matter. Such an appreciation forms in man an attitude of awe, wonder and respect for the diversity and richness of creation. Furthermore, when it is appreciated that the natural world does not exist solely for man's purposes, but for God's, man's response is one of humility and reverence (rather than of exploitation and arrogance). Because the world is God-created, it must be approached with an attitude of reverence and respect. Activities which are not characterised by this attitude or which involve the destruction of parts of this variety and the plurality of God's creation (particularly species and habitat destruction) may be regarded as inconsistent with this attitude' (Russell, 1983:9).

4.4   Discovering what animals think and feel - the role of ethology and biochemistry.

It has already been established that appeals to anthropomorphism and sentimentality distort the debate on the human/animal relationship. In the past, however, such failings were difficult to avoid, as it was only in examining the similarities between animal and human experience and by establishing consequential analogous links that we could begin to understand and appreciate what their feelings might be. As a result, it was easy to transfer our perceptions and emotions onto other species and to dress them with human characteristics that denied them any essential dignity that they might have possessed. This clearly contradicts the attitude exemplified by Russell (1983) and others whereby the member of another species is respected for what it is as a cow, sheep, pig or whatever, rather than as surrogate human.

Indeed, Midgely (1984:142) describes anthropomorphism as 'a special, fallacious and misleading way of reasoning, displayed whenever people attribute either conscious experience in general or particular conscious states to animals. Its fault consists in the extrapolation of notions derived from human experience to apply to the non-human'. In a similar vein, Kennedy (1992:167) dismisses our tendency to indulge in imaginative anthropomorphism towards animals as a disease with which we've been 'genetically programmed' and 'culturally inoculated'! Although there is undoubtedly some truth in this, which is far from being confined to a by-gone age, it does not completely invalidate the argument from analogy. As long as it is combined with other available sources of evidence concerning our understanding of how animals perceive the world around them, the argument can still perform a useful if limited function.

The Report from the Brambell Committee (1965:9) recognises the value of an empathetic approach towards other species. 'The evaluation of the feelings of an animal must rest on analogy with our own and must be derived from observation of the cries, expression, reactions, behaviour, health and productivity of the animal.....our understanding of their feelings is not different in kind, but rather in degree, from that which we form of a fellow human being'. In placing such an emphasis on identifying the similarities between human and animal feelings, Brambell is in accord with Scruton's emphasis in the crucial role of sympathy in all moral relationships.

Whilst recognising the dangers inherent in this approach, Dawkins (1980) suggests that, once all the material concerning the feelings of other species has been gathered together from the widest possible range of sources, it can still only be fully assessed by our applying the argument from analogy. Only then can moral decisions be made. 'Such an assessment will inevitably rely on recognition of some similarity between ourselves and other species to make the final leap between what is observed directly by us and what is experienced subjectively by them' (Dawkins, 1988:98).

This leads her to appeal for as accurate an understanding of the similarities between human and animal experiences as can be achieved and for as little use of the imagination as is possible. She then vividly illustrates the consequences of ignoring this advice and the confusion that arises out of applying an exaggerated anthropomorphic imagination to the subject of animal welfare by citing the example of the tapeworm. 'It would be a mistake to assess the welfare of tapeworms by trying to answer the question "How would you like to live inside someone's intestines?" Their "peptic Nirvana", as Julian Huxley once called it, has nothing to do with the conditions that are best for human welfare or vice versa' (Dawkins, 1988:101).

In passing, it is important to recall what was discussed earlier, that until comparatively recently the only alternative to the subjective identification of similarities between human and animal experiences appeared to be the adoption of the completely opposite point of view. Their experiences and ours were deemed to be so different that humanity was justified in ignoring any claim that they might have on our sympathy or concern. Descartes' thinking took this to the extreme by reducing the status of an animal to that of a machine. Comparing their physiological complexity to the intricate workings of a clock, he dismissed their cries under duress as being purely instinctive. The sounds have no meaning and should not be interpreted as indicating that the animal is in distress. Rather Descartes suggests that such cries should be likened to the sound emitted by a clock's spring that has been touched (Singer, 1995:200). Such a doctrine was bound to have many disturbing practical consequences, as is cruelly illustrated by those eighteenth century vivisectionists who, with a clear conscience, dissected living unanaesthetised animals. Believing that their victims were machines that were purely governed by instinct, devoid of any sensation including pain, they consequently assumed that they could treat them without concern or regard (Regan, 1988:5).

Clearly, just as the argument from analogy is to be rejected on the grounds of inadequacy when it is proposed as the sole means of understanding an animal's perceptions and feelings; so too Descartes' thinking is to be rejected as an alternative, on the basis that it is neither ethically nor scientifically acceptable. As one depends on imprecise, subjective perceptions, so the other fails to acknowledge the possibility that the animals concerned might actually possess needs, sensations and feelings. Both deny those creatures any sense of respect for what they are. Fortunately, however, the choice is not limited to these two schools of thought. Recent decades have witnessed the emergence of a wide range of new methods for establishing a more precise and objective understanding of how other beings experience and view the world. As Fraser and Broom (1997) have indicated, this has come about, 'as people with backgrounds in zoology, physiology, psychology, animal production and veterinary medicine have investigated the effects of different conditions on animals' (Fraser and Broom, 1997:4).

Dawkins (1988:2) has acknowledged the emergence of this useful range of academic and scientific resources, although she has then argued that 'each one is inadequate on its own. What is necessary is a synthesis of the pictures given by all methods'. Indeed, it is only this synthesis that can achieve the goal which has long been felt desirable but has always seemed beyond human reach, of allowing animals as far as is possible to 'speak for themselves' (Webster, 1994:6), i.e. to inform us of their preferences, needs and sensations. Furthermore, whilst our understanding of what is being communicated to us may as yet be far from perfect, it is now possible, as a result of our ability to assess and measure an animal's welfare, to establish practical principles for expressing respect towards them.

With regard to farm animals, awareness of their needs and preferences resulted in the establishing of basic welfare criteria by the Farm Animal Welfare Council (FAWC) in terms of the 'Five Freedoms' (page 28), the implications of which will be examined in the next chapter. As a result of these developments it is now possible to make better-informed decisions concerning the legislation that affects the welfare of all species of animal life including those employed in farming systems. The formulation of that legislation can now be based on scientific evidence rather than pure sentiment or imaginative conjecture. Indeed, recent debates concerning certain livestock production systems and the legislation that has been a consequence of those debates illustrates this. To quote two examples, the regulations governing the transportation of animals and the imminent phasing out of battery cages in the poultry sector have both been guided and informed by the work of ethologists and biochemists alike.

Webster's idealistic picture of animals now being able to 'speak for themselves', in that they are able to indicate their preferences, is in fact illustrated in both of the instances quoted above. Some understanding of animals' feelings and perceptions has been accomplished without having to resort to any analogous comparison with human experience. Instead, indices for determining and measuring animal welfare have been applied to the studies, indices that Eddison (1995:388-391) categorises into four groups: 'physiological, biochemical, production and behavioural.'

In assessing the welfare of an animal, it is easy to make certain misplaced assumptions if only one of these categories is used. For example, whilst Eddison identifies production factors as an indicator of welfare, it would be wrong to assume that an animal which is healthy, fertile or efficiently productive (e.g. in terms of weight gain) is therefore, by definition, content. Dawkins (1988:29) illustrates the fallacy of using such criteria as the sole means for measuring welfare by quoting the example of productive, healthy turkeys kept in environments that raise serious welfare questions. Their weight gain, for example, can be achieved by restricting their exercise and their healthy condition can be maintained by the continuous application of prophylactic drugs rather than the provision of a healthy environment.

This is not to dismiss observations of an animal's state of health as being worthless, which would be ridiculous, but to emphasise again the importance of weighing one assessment against another. Indeed, Eddison appreciates the problems but still emphasises the importance of taking an animal's productivity, and therefore health, into account. 'If individuals do not grow at the normal rate, fail to come into oestrus, or have reduced conception rates or litter sizes, we can probably conclude that they are not in a state of good welfare' (Eddison, 1995:388-391). This, he then suggests, is frequently because of illness or some other health factor.

With regard to the physiological and biochemical categories that Eddison identifies, Fraser and Broom (1997:267) have defined how changes in these areas can be assessed and measured. 'Physiological responses to difficult conditions include orientation, regulatory responses, and preparation for flight or defence. Changes which can be measured include those in heart rate, ventilation rate, adrenal functioning and brain chemistry'. Such measurements provide a precise picture of the effect of circumstances or environment on particular animals. The investigations carried out by Fordham *et al* (1989:103) for example into the stress levels experienced by lambs at an abattoir illustrate how valuable such an assessment can be. The aim of the study 'was to examine the release of β-endorphin and cortisol in lambs after routine stressful stimuli such as handling, transport and slaughter at a commercial abattoir.' The results indicate, for example, a significant increase in the concentration of plasma β-endorphin in the blood samples taken from animals that have experienced stressful situations compared with those in an unstressed state. As a result, it is possible to arrive at a precise, measured determination of the stress levels being experienced by the lambs and to know in effect what they are feeling.

Eddison's final category of welfare focuses on the observation of animal behaviour, an area of study, which he suggests, is 'fundamental to welfare assessment.' This is because animals are 'highly motivated to perform certain behaviours (e.g. nest-building by pre-parturient sows or pecking for food by hens)' (Eddison, 1995:388-391). The frustration of these instincts will lead to abnormal behaviour, which Eddison describes by drawing on Fraser and Broom's definition and categorisation. 'They catalogued four main groups: stereotypic or repetitive, non-functional behaviour; behaviour directed towards self or inanimate objects; behaviour directed towards other individuals; and abnormal function' (Eddison, 1995). The study of such behavioural patterns and a consequential understanding of the reasons for them provide a major source of evidence for assessing an animal's state of welfare. Each display of abnormal behaviour communicates something of that creature's frustration, anger or fear.

Further understanding of an animal's perception of its environment by way of observing that animal's behaviour can be achieved by studying the choices it makes and the energy it is willing to expend on attaining its preferences. Known as the "economic theory", it enables us to discover what matters to an animal by allowing that animal to put a price on what it wants to choose in terms of the effort it is willing to make to achieve it. As a result it is possible to discover what the desires and wants, and especially the priorities and preferences of any particular creature may be. So for example, Bubier (cited by Dawkins 1993:155) examined how hens value dust-bathing and scratching for food. There had long been evidence of battery hens displaying such behavioural traits even when their sterile environments denied them the access to the actual means of fulfilling their desires. What was not known was the degree of frustration and distress that this was imposing on the hens concerned. By arranging access to different floor types, reached through different size openings and various difficult courses that had to be negotiated, it was possible to discover how much effort the hen was willing to put into attaining the desired environment. The results not only showed that poultry placed great value on a practice such as dust bathing but that they also experienced great frustration when that practice was denied them.

As has already been recognised, each of these methods for measuring the degree of stress or welfare being experienced by an animal needs to be applied and assessed in conjunction with one or more of the alternative indicators. This cross-dependence of methods has been emphasised by both Dawkins and Eddison. Whilst acknowledging, for example, that physiological and biochemical changes in an animal are a valuable source of evidence for obtaining an understanding of how that animal is experiencing a particular situation or environment, Eddison urged caution in interpreting what those changes mean. He illustrated his concerns by citing the cardiovascular changes experienced by sheep being loaded or transported and advises that 'care must be taken in interpreting such results since normal activities, such as play, running, courtship and mating will also elicit similar results' (Eddison, 1995:388-391). Similarly, in using a simple check list which was first formulated by Morton and Griffith in 1985, he was not only able to establish criteria for making objective judgements regarding an animal's state of welfare but was also able to indicate how those criteria frequently relate to each other. In identifying cardiovascular changes, for example, as one of the indices for measuring animal welfare, and placing it alongside other recognised indices such as alterations in appearance and behaviour or changes in eating and drinking patterns, he was able to establish a system by which an overall assessment of welfare was achievable.

By methods such as these the experiences of a non-human can be known and understood and from that knowledge base responsible decisions can be taken about the treatment, use and welfare of animals. This in turn means that it is now possible to establish which livestock production systems are in accord with the moral and theological principle of respecting an animal for what it is and those that are not.

4.5 An expression of concern - the role of the animal protection organisations.

The insights that recent scientific developments have given us into the perceptual world of non-human species have clearly been of great benefit to all who work towards the goal of improved animal welfare. This has been particularly true for those entrusted with the formulating of relevant legislation. Equally, those who seek to influence the legislative process by lobbying those responsible have also made use of these same insights. Recent decades have witnessed a burgeoning of such organisations committed to the cause of lobbying on behalf of animals. Indeed, the majority of the animal protection groups listed by Clough and Kew (1993:173 ff) have been founded since the 1960s. Many of these have retained a commitment to the general welfarist stance of long-established organisations such as the RSPCA, whilst others have adopted a more strident and radical policy.

Largely inspired by the philosophy of animal rights, the goal of such organisations tends to be the eventual eradication of any practice which they believe abuses animals, regardless of whether the perpetrators are farmers, scientists, entertainers, sportsmen or pet owners. The strength of feeling evoked by such views has in some instances led to divisions within the animal protection movement itself, as those welfarists who have engaged in dialogue with those who use animals, have been accused of compromising with the enemy (Ewbank, 1994:5). The most extreme of these organisations have advocated and carried out violent attacks, not only on property and premises linked to practices deemed to be abusive of animals but also on the individuals concerned. As the people and groups who engage in such acts tend to place themselves above the law and appear to have little respect for opinions different from their own, they consequently remove themselves from any forum for debate which tries to establish compatibility and conciliation. It would seem that their relationship with the livestock industry would only ever be one of antagonism and hate.

Whilst such violent protests against the use of animals for human purposes is usually restricted to a small and extreme minority, that minority is dangerous and well organised and as a result cannot be ignored. Their activities have been well documented on many occasions by the media coverage of organisations such as the Animal Liberation Front (ALF). In August 1997, for example, Jenkins ('The Times' 30.8.97:6) reported court proceedings against five ALF members who 'devoted their lives to inciting people to violent action, from burning lorries to sending explosive devices through the post'. The organisation of their programme of terrorism extended to the publication of instructions for the making of bombs and incendiary

123

devices in specialist magazines and books and the using of the Internet to recruit other like-minded individuals to their cause.

In another incident reported in 'The Farmers Weekly' (18.9.98:12) a lorry driver was critically injured by rights activists who dropped a rock on his lorry cab in the belief that his vehicle was carrying livestock. These actions were subsequently criticised by representatives of animal welfare groups who distanced themselves from such behaviour by describing it as 'sickening' and 'counter-productive'. Despite such criticism however, it must be recognised that such violence is endemic in certain extreme elements of the animal rights movement, and is a fundamental tenet of their philosophy. In describing the aims and objectives of the Animal Liberation Front for example, Clough and Kew (1993:176) state that these include the 'destruction of property of animal abusers' and 'economic sabotage', neither of which are conducive to objective debate.

Fortunately, the ALF does not represent the majority of animal protectionists, who in the main deplore the use of physical violence as a way of promoting their beliefs. In a feature in 'Agscene' (Summer 1997:19), for example, Penman calls upon his fellow rights activists to engage in non-violent protest against what he emotively describes as 'crimes committed against animals and the environment.' He continues, 'If people want a peaceful and just society then we have to embrace these qualities both in our personal lives and in our campaigns.' In a similar vein Clough and Kew (1993:xiv) bemoan the frequently projected media image of the 'balaclava-clad, lentil-munching arsonist' style of animal rights activism as a hindrance to the promotion of animal issues.

Such denouncements of the vociferous and violent element within the rights movement should encourage those involved in livestock production and similar activities to recognise that reasonable people can be found on both sides of the debate. Consequently, it should be possible to achieve some degree of mutual understanding and respect along with the added possibility of accomplishing gains not only for animals, but also for those who work for them and those who work with them.

In the main, whilst they may remain opposed in principle to many production methods and systems, most of the animal protection organisations are willing to work with the farming community towards improving welfare standards in the industry. Compassion in World Farming (CIWF) epitomises such an approach. The organisation has been a vocal opponent of many past as well as current agricultural practices, from livestock transportation to veal production and from sow tethers to battery systems for poultry. Well-organised and highly skilled in public relations and the use of the media, it would seem to have a greater impact on public perceptions than any other welfare group. Indeed, it describes itself as 'the leading organisation campaigning for an end to cruel factory farming systems' ('Agscene' Spring 1995:2) and is consequently viewed from within the agricultural industry as being 'anti-farmer'.

What is clear from many of the articles in their magazine 'Agscene' is that a large number of their supporters are advocates of animal rights, and consequently are also committed vegetarians or vegans. Nevertheless, none of this prevents them from meeting with representatives of the farming community in an attempt to resolve some of the issues that they find contentious. Occasions such as the consultation held in January 1995 between representatives of CIWF and the NFU to discuss the export of UK livestock for slaughter in other countries illustrate a practice that needs to be encouraged ('Agscene' Spring 1995:2). As Clough and Kew (1993:45) note 'CIWF while encouraging the move towards the animal rights ethic generally have the shorter-term objective of improving present conditions for livestock.'

For those representing farming, meetings with groups such as CIWF are bound to be difficult, as they know that the people with whom they meet will criticise and deplore the way that they earn their living. It is essential, however, that this consultative process is encouraged, as it is only through dialogue that any sort of agreement can be reached and any degree of understanding can be achieved. As will be illustrated in the next chapter, UK farmers rightly express pride in the high standards of welfare and stockmanship that characterise their livestock industry. The responsible attitude they exhibit towards the needs of animals has to be communicated to those who at present neither appreciate nor understand it.

Furthermore, a lifetime spent in the company of livestock farmers has convinced me that the vast majority do care deeply about their stock. This too needs to be conveyed to those who may currently believe that all stockmen are animal abusers with no sympathy or concern for the creatures in their care. How far this is a distortion of the truth is frequently illustrated in angry articles and letters in the farming press condemning incidents of unmitigated cruelty towards livestock and those who perpetrate them.

In an article in 'The Dairy Farmer' (1.9.97:12), for example, Ginever, who is himself a dairy farmer, declared his revulsion at some of the implications of the calf slaughter policy in parts of the European Union. Having watched a television programme on the subject, he expressed his anger at the way that the calves were handled and the suffering they had to endure. His practical experience of and belief in the highest standards of stockmanship caused him to view such treatment of animals as being grossly offensive. I suggest that most livestock farmers would agree. In this, they are at one with those who would normally criticise anything to do with animal production. Consequently, it would seem that there should be the possibility of further respect, dialogue and understanding between the two sides in the debate concerning the well-being of farm animals.

In addition, if as has been suggested, the principle of respect is to be the governing ethic for livestock production, then logic surely dictates that that same respect should be extended towards members of our own species even when we disagree with them. As has been recognised, debate must be encouraged if confrontationalism is to be overcome, and this can only happen in an environment in which each side respects the views and integrity of the other. In fact, the establishing of a forum for public debate that draws together representatives of animal rights and welfare groups on the one hand, and livestock farming interests on the other is long overdue. One can only hope that a university or public body will eventually take the initiative to create such a forum. Indeed it seems from the aims of some animal protection organisations that they already recognise the need for such a forum and are attempting to respond to it. Through its Fellowship, the International Fund for Animal Welfare (IFAW), for example, is encouraging research into 'competing theories of moral obligation to animals.' Another project is engaging in a 'critical analysis of the concept of animal rights'. Both projects are in line with one of the organisation's stated aims, which is 'to provide a basis for informed and critical debate'. In particular the IFAW Fellowship will encourage and facilitate, wherever possible, dialogue with other disciplines especially with those engaged in practical work with animals' (IFAW, 1996:8-9).

Sadly, whilst the aims are admirable and the research projects promising, both are apparently dominated by the entrenched acceptance of the principle of animal rights, which seems to be an integral and over-riding feature of the IFAW philosophy. Consequently, it appears that little attention or credibility is given to opposing views, including those of the livestock producers. This is not to belittle or condemn IFAW's work, but to hope that genuine and critical debate in an open forum will be accomplished and that sympathy for and understanding of the livestock producer's views will be achieved. When and if this happens IFAW could well be the body to establish the forum which would gather together all parties interested in animal issues. IFAW's desire to promote research into issues concerning animals is shared by a number of other welfare bodies, including the Universities Federation for Animal Welfare (UFAW) and the Humane Slaughter Association (HSA). Whereas IFAW concentrates on the philosophical and theological aspect of the debate, however, both UFAW and HSA are concerned with the scientific and technical context. Moreover, whilst the philosophy of animal rights is a major influence on IFAW's strategy and stance, the work of both UFAW and HSA are based on the principle of human responsibility towards other species.

For example, UFAW, which was founded in 1926, describes itself as a 'scientific and educational charity concerned with the welfare of animals in situations where they are affected by human activity or dependent on people. UFAW does not engage on either side in public controversies concerning the use of animals and on the whole does not make judgements... UFAW concentrates on improving the welfare of animals that *are* used - a matter which is sometimes overlooked in the debate on whether or not we should use animals' (UFAW, 1995). In a later definition of the organisation's aims (UFAW, Spring 1996) the means for achieving the stated goal are outlined. They involve the examination of an animal's physical and psychological needs and a 'working towards better standards of care, reducing any pain or distress and ensuring that when they have to be killed, they are killed humanely.'

UFAW's aims, as well as its willingness to work alongside the practical farming community, are reflected in the variety of research projects that the organisation funds. A sample list of these (UFAW, 1996) includes work into the welfare implications of outdoor management systems for sows, the development of a big bale straw dispenser for pigs, the overcoming of problems regarding the provision of perches for hens in laying cages, and an examination of the problems of hoof lesions in lame dairy cattle. All of these are clearly of economic as well as moral relevance to livestock producers, and indicate how an organisation dedicated to improved welfare standards for animals can be compatible with good farming practice and economic need. Indeed, the principles on which UFAW is based are very much in line with the ethical stance already proposed as the foundation for responsible livestock production.

Similar comments can be made about the work of the HSA. Founded in 1911 as The Council of Justice to Animals and Humane Slaughter Association, it is 'the only charity committed to the welfare of animals in markets, during transport and to the point of slaughter. It seeks to achieve its objectives by a positive, practical approach and by maintaining a close working relationship with those responsible for the slaughter, marketing and transport of livestock' (HSA, 1996:2). Although the organisation's terms of reference are very specialised, they are clearly of great relevance to the livestock industry, especially as that specialism gives them an expertise that is unique. It was as a result of the HSA's efforts for example, that in the 1920s the pole-axe was replaced by the captive-bolt pistol as the means of slaughtering animals in the UK. A more recent example concerns amendments to UK firearms legislation which were passed in February 1998, allowing people to own captive-bolt equipment without having to hold a firearms certificate. At first sight, this would seem to be an improvement regarding the humane treatment of injured stock. Any farmer could now possess the means of despatching such stock without having to cause the animal any further distress by awaiting the arrival of a vet or licensed slaughterman. As the HSA were to point out, however, a captive-bolt stunner in the hands of someone who is not competent to use it can cause further distress and suffering to an animal, rather than relieving it. Consequently, whilst a firearms certificate is no longer required by an individual possessing such an item of equipment, a slaughterman's licence is. In reporting on this particular incident, the HSA (1998:2) remarked that they are 'the only animal welfare organisation with the specialist knowledge to deal with these important issues. Without the input of the Association, it is probable that many of these problems would not have been addressed.'

There are two further areas in which the HSA have probably had their greatest influence. The first relates to their practical development of the provision of mobile slaughterhouse facilities. The benefits in terms of animal welfare are obvious, in that the taking of such a facility onto a farm or into a central location reduces the stress of transportation on the stock to be slaughtered. In addition, it provides 'a humane, hygienic service for remote or rural areas, specialist species (e.g. deer), routine culls and developing countries' (HSA, 1996:9). The project has faced many difficulties, not the least being the severe regulatory restrictions that are imposed on all abattoirs and which have already brought about the demise of many small rural slaughter houses. As the two mobile units become established, however, they should provide real welfare benefits for stock in the areas in which they operate and should provide a model that will be copied in other parts of the country.

The second area in which the HSA has had a major impact, and which is of particular relevance to livestock production, is that of training. Recognising the absence of any established standards for assessing the skills of those who work in areas such as slaughter houses, auction marts, and livestock transportation, HSA have responded by initiating courses and qualifications for each discipline. The implications of these measures in terms of improved standards of livestock handling and therefore of animal welfare are obvious, and again illustrate the benefits for all concerned of practical co-operation between the welfare organisations and the farming community.

Whilst the efforts of the HSA have brought about many improvements in the welfare of animals in such places as the abattoir and the auction mart, other organisations have also played a part in achieving these and related improvements. Notable amongst them is the RSPCA. The earliest of all the animal protection bodies in the UK, it has always had close involvement with the farming industry, fulfilling a policing and monitoring role with regard to many of the industry's activities. In 1824 for example, at his own expense, the founder of the Society employed an inspector to monitor the treatment of stock at Smithfield Market. This established a practice that continues today and currently extends to all livestock marts in the UK.

Although the RSPCA has been critical of certain livestock production methods and has always been opposed to intensive systems (RSPCA, 1994:10), it has still endeavoured to work with the livestock industry in an effort to improve the welfare of farm animals. One of the most impressive results of this working relationship has been the establishment in 1996 of Freedom Foods Ltd. Described by the Society as 'the most exciting development in farm animal welfare for many years' it is a labelling system for animal based food products that 'recognises good farming practices and provides the assurance consumers now demand for quality products produced to high welfare standards' (RSPCA, undated). Both the name and the welfare standards that are imposed under the scheme are taken from the Farm Animal Welfare Council's model of the five freedoms (see Section 2.5.1), each of which is considered to be an essential element in any acceptable system of livestock production. Although some producers have been suspicious and at first resented the thought of being 'policed' ('Farmers Weekly' 8.5.98:5), the scheme has now attracted a great deal of support from producer and consumer alike. The Society has reported 'growing retail demand for Freedom Food labelled products as well as an increased public awareness and interest in the scheme' (RSPCA, 1997).

Farming companies have also seen the value of the system, but in economic as well as welfare terms. This is illustrated by the Long

Clawson Dairy for example, which in producing a Blue Stilton with the Freedom Foods label, became the first dairy company to become involved in the scheme. Not only was the Dairy aware of the ethical principles that the Freedom Foods label upholds but it also recognised how the label can be a valuable asset in marketing the cheese to those concerned about other production systems ('Farmers Weekly' 8.5.98:5). Similarly, a number of egg producers have realised and subsequently utilised the marketing potential of the Freedom Foods welfare standards. Freshlay, for example, now advertise their 'Granary Egg' as being 'produced under welfare and food safety conditions and practices as laid down and monitored by RSPCA Freedom Food and the British Egg Industry Council Lion schemes. The egg is traceable back to its farm of origin' (Freshlay, undated).

The symbiotic partnership that is represented by Freedom Foods Ltd., between an animal welfare organisation and the farming community, provides a model that other organisations would do well to copy. It has proved that improved standards of animal welfare in agriculture can be profitable for the farmer as well as being beneficial to the stock. In addition, it has taken account of the experience, knowledge and skill of those who regularly work with livestock and has listened to what they have had to say ('Farmers Weekly' 8.5.98:5). The RSPCA are to be commended for taking such a lead in this field and it is hoped that other welfare bodies will follow.

Whilst there are many other groups committed to improving the conditions under which animals are farmed in the UK, those mentioned above give an indication of the range of concerns, beliefs and strategies that are represented. Some are clearly sympathetic to the world of agriculture and are thus willing to work with farmers in trying to achieve their goals. As groups such as UFAW and HSA tend to uphold the farmer's ethic of responsible stewardship and respect towards other species, it is possible for them to establish a working partnership with the farming community. The ethical stance adopted by some other organisations, however, may well create difficulties in forming such a partnership, especially when that stance is firmly based on a belief in the concept of animal rights. Despite this, I believe that it is still possible to achieve a degree of understanding and respect when there is mutual openness to the possibility of engaging in debate.

There then remains the third category of animal protectionist, that of the animal rights extremist. It would seem that the possibility of engaging members of groups such as the Animal Liberation Front in any form of meaningful or reasoned debate is remote, dedicated, as they are, to achieving their goals at any cost, and ready to resort to violence as a means of achieving their ends

4.6 Influencing diet

There are two further groups which currently exert a considerable influence over the public's perception of matters relating to livestock production, and whose role must therefore be taken into account. Although their interests are diametrically opposed to each other, each one is concerned with the place that meat holds in the human diet. On the one side of the divide, there are the meat retailers who have an obvious interest in ensuring that people continue to eat and therefore purchase the product that they sell. On the other, there are the vegetarians who, in trying to achieve their vision of a society in which meat has no place on the menu, want to see no one buying meat at all.

As both groups claim to have strengthened their influence in British society over recent years and as they both express an interest in animal welfare, they clearly have a place in the current debate concerning the ethical base for intensive livestock production. In addition, as they both have the ability, desire and commitment to exercise and extend their influence even further - one for the purposes of financial gain, the other largely for reasons of principle - it would be folly to ignore what they say and do.

4.6.1 The meat market

As will be illustrated below, supermarkets and catering establishments have dramatically increased their share of the retail food market in the United Kingdom over recent decades. A consequence of this has been the increased ability of both groups to mould and direct the buying habits of the consumer. This in turn has had an impact on the marketing of livestock products, as well as on the methods employed in their production.

The power that the supermarkets are able to exercise over both the livestock industry and the consumer can be seen in the statistics for meat sales in the United Kingdom. These indicate, for example, that 70% of the meat purchased for domestic consumption in this country during 1997 was bought across a supermarket counter (see Table 6). This increased to 77.4% for the purchase of pig meat (MLC, 1998:42) and 72.5% for poultry products (MLC, 1998:50). Quite simply, supermarkets effectively control the meat trade in Great Britain.

Table 6   **Percentage of total retail meat sales by source**
(From MLC, 1998:58)

Sources of total meat purchases	1996	1997
Independent butchers	17.7	16.2
Co-operative Wholesale Societies	2.5	2.2
Supermarkets	67.5	69.9
Independent grocers	1.2	1.0
Freezer centres	5.3	5.2
Other retail outlets	5.8	5.5
Total	100.0	100.0

The diet and tastes of the consumer are also increasingly being influenced by the growing popularity of eating in restaurants and fast-food outlets (National Farmers Union, 1998:40). Between 1985 and 1993 consumption spending in UK cafes, restaurants and hotels increased by 50% (National Farmers Union, 1998:47). Not only is this set to continue, but it is reaching the point where it seems likely that most meat-based meals will soon be consumed outside of the home (National Farmers Union, 1998:44). It follows, therefore, that the influence of a company such as McDonald's has to be of great interest and concern to UK livestock producers, especially as they boast of feeding 2.5 million people every day in Britain alone (McDonald's, July 1998). This interest will be further encouraged by the company's pledge to source the major part of the supplies they need for their British operations from local producers.

The larger the share of the market controlled by any one company, however, the more power and influence that company has to direct and control the policy and production methods of the farms that supply them. This is clearly illustrated by the document from which the above figures concerning McDonald's sales are taken (McDonald's, July 1998). In it the company not only state their commitment to British agriculture as their main supply source in the UK, but they then proceed to emphasise the high standards of animal welfare they demand from those suppliers. In this particular case, the retailer's influence would appear to be beneficial to everyone concerned - consumer, producer and animal alike. The consumer is assured of a safe, traceable, high-quality product, the producer is guaranteed a regular outlet for his meat, and the animal is reared in a welfare friendly system.

The problem arises, however, when one company or a small group of companies having taken control of a major share of the market then moves from the sphere of influence into that of control and exploitation. Regarding the consumer, such control is achieved by using skilful advertising and marketing to sway what he/she wants and subsequently buys within a choice-limited market. With regard to the producer, it is accomplished by restricting the size of the market place in which they are able or allowed to operate, to the point where they have few alternative options regarding potential buyers. Such limitations easily lead to the abuse of 'retailer power' and the potential for the making of further and tighter demands on the supplier with regard to methods of production. As these demands are likely to be motivated by a drive for greater economy in production costs and larger profit margins over the counter, the net results are potentially detrimental to the livelihood of the producer as well as the welfare of the livestock.

In November 1999 'Farmers Weekly' (19.11.99:7) reported an example of such retailer power. The Safeway chain of supermarkets had launched a promotional campaign which guaranteed 'customers that the products they buy the most, such as fresh fruit, vegetables and meat are always on the shelves.' At the same time the supermarket's suppliers had received letters informing them that they would be invoiced £20,000 per product line as their contribution towards the campaign ('Farmers Weekly' 19.11.99:7).

For some suppliers the potential cost was hundreds of thousands of pounds with the fear that failure to pay would prompt commercial reprisals. For farm businesses already facing the worst financial crisis in UK agriculture for sixty years there was little chance of finding an alternative buyer for their products. Equally there were few ways left open to them for cutting costs further than they were doing. The net result, therefore, was increased financial anxiety for the producer and the likelihood of deteriorating standards of product control and possibly animal welfare as suppliers were forced to cut corners even further.

Further concerns over the perceived power of food retailers was expressed in a plea from the NFU (March 2000:1) for farmers who had experienced problems in supplying the supermarkets to provide the Competition Commission's on-going enquiry into retailer power with relevant details. The article commented that farmers' fears over 'de-listing' was discouraging them from submitting evidence. Such fears have not only been expressed to me personally, but have been supported by a letter from one supermarket company to its suppliers threatening de-listing to anyone having dealings with certain suppliers. A copy of that letter was passed on to the Minister of Agriculture for action. I believe that the exercising of retailer power in this way is unacceptable and is effectively a form of blackmail.

Concerns such as these have been voiced in the USA for a number of years over the behaviour of a small number of powerful companies increasingly monopolising the food industry. It would seem that the control of basic resources, agricultural production, transport, and retail outlets now rests in the hands of an ever-diminishing number of players. This was recognised by Heffernan (1995:17), who stated that 'the control of the production and processing of most of the nation's food has become concentrated in the hands of a few transnational corporations (TNCs) with dire consequences for rural America.' Having identified those companies, she continued, 'Three or four firms, most often TNCs, now control between 40 and 80 percent of the slaughtering, milling and processing and shipping of most grains and livestock in the United States.' The concern expressed in the paper was for the welfare of rural communities, especially those engaged in farming, when faced with the potential bullying and exploitative powers of these companies.

Similar fears exist, however, over environmental and animal welfare issues, when their well-being conflicts with the over-riding interests of these companies. Indeed, something of this was illustrated in the environmental concerns arising out of Monsanto's efforts to impose the technology of genetically modified crop seed and food products into Britain. 'Monsanto, whose GM food sales will soon be worth £8 billion a year, is able to exploit its position in the market, say these critics, to create dependency on its products' ('The Week' 20.2.99:9). When power of this magnitude is exercised in the agricultural market place, it is right that all concerned about welfare, be it of the environment, the consumer, the livestock producer or the actual livestock, should be anxious regarding the future.

Such a scenario was illustrated to some extent in 1998/99, when a number of meat retailers were heavily criticised for gross profiteering during the crisis that hit UK pig producers that year. Despite their open protestations of support for welfare friendly production systems, some retail companies were importing cheap supplies of meat from foreign competitors who were using methods of rearing and slaughtering that are illegal in this country. To many, it appeared that the supermarkets' domination of the meat retail trade was unethical, and that they were accountable to nobody but their shareholders. An article in 'Pig World' expressed something of the anger that pig farmers felt at the time regarding this display of arrogance and hypocrisy. 'Pig farmers have had faith in the retail outlets who asked for (or was it demanded?) superior welfare, the highest in history. That they now feel rejected and dejected is understandable. Betrayal is perhaps a more appropriate word' (Walton, 'Pig World' Sept. 1998:5).

Another article in the same magazine similarly reflected the feelings of many producers. 'UK pig-producers have moved to high animal welfare and food safety systems in response to the wishes of consumers and government. We have unilateral legislation to ensure our industry operates to the expected standards... So now it is the responsibility, indeed the moral obligation, of every retailer to ensure all products sold in this country comply with the same standards demanded by the UK government' ('Pig World' Sept. 1998:47).

Having worked closely with the Pig Industry Support Group during the crisis that the industry experienced in 1998/99 I have a great deal of sympathy with those who believe that government, consumer and retailer betrayed them and who have seen their farms and businesses collapse. Indeed, I believe that their criticism of both supermarkets and government is largely justified in that cheaply-produced pigmeat from systems, that on welfare grounds would be illegal in this country, was allowed free entry into the UK and as a consequence destroyed our own pig industry.

One wonders, however, how far the recent and expensive welfare developments in UK pig production systems have been introduced onto British farms purely as a result of legislative requirements. Indeed some might even suggest that the reference to 'high' animal welfare systems is a misnomer, and that the systems and standards now in place are a base line from which there is room for continued improvement. Whilst I would like to believe that the changes took place as a result of a moral conviction amongst farmers that previous systems and methods were unacceptable, I am not sure how far this was the case. Given the current economic climate in livestock farming it seems improbable that many UK farmers would have made the required changes to their methods of production. What remains unacceptable, however, is that, having made the required changes, British pig producers are still being effectively penalised, as they are forced to produce pig meat in expensive welfare-friendly systems whilst retailers are still allowed to sell cheaply produced imports from systems which are illegal in the UK.

Pig producers have not been alone in dealing with the supermarkets' double standards, as representatives of the poultry industry have been only too willing to point out. Delegates to the conference of the Rural, Agricultural and Allied Workers (RAAW) Trade Group, which took place in June 1995, criticised the hypocrisy of supermarket buying policies on matters of animal welfare. Scott (1998) said that 'supermarkets were guilty of double standards when dealing with poultry suppliers in the UK against their overseas competitors... "They demand the strictest hygiene and welfare standards from UK producers. Yet they are willing to import, in our opinion, inferior products from the EU and beyond, simply because it is cheaper."' Another delegate commented that 'the producers are being squeezed by the big supermarkets, to reduce the price of products to them. The supermarket protocols are a sign of their hypocrisy - they demand the strictest welfare yet still want products on the cheap and fail to recognise the problems they cause' ('Landworker' July 1998:5).

Such views are not only representative of feelings being expressed across the livestock industry, but they are also heard from the representatives of the government. When, for example, Elliott Morley, the minister responsible for animal welfare, addressed the UK 'People in Pigs' Conference in November 1998, he stated that 'the industry needs to use the issue of welfare to their advantage. We have an excellent product to sell. British pig producers have put a great deal of effort and a great many resources into raising welfare standards and ending the sow stall and tether system. In addition, pigs in this country are not fed bonemeal, as they often are on the continent. These standards of welfare must be recognised and reflected in the way that the products are marketed' (MAFF 11.11.98).

This was followed by the report that the British Retail Consortium, the representative body for all the major UK supermarkets, had agreed that as from 1st. January 1999 'all fresh pork and bacon sold in their outlets will come from stall and tether free systems and from pigs which have not been fed bonemeal.' From this, it was assumed that the retailers were making a united moral stand for welfare friendly production systems, and a commitment to only sourcing their meat from suppliers who complied with such systems.

In the February 1999 edition of 'Pig World', however, concern was expressed that some were still failing to abide by the agreement, with the Morrison and Iceland retails chains coming in for particular criticism. The impression created was that in some parts of the retail food trade animal welfare concerns were only relevant when they had no adverse effect on profits or alternatively when they could be used as a marketable commodity.

Fortunately, such accusations cannot be directed at all of the supermarkets. Several of the leading companies, for example, have expressed strong support for those farm assurance schemes that have encouraged improved standards of health and welfare for livestock. This support has then been further endorsed by the products from such systems being given a prominent place in the stores. Indeed, some companies have taken their commitment to the development of welfare-friendly systems even further by introducing their own schemes and standards.

One company that exemplifies such an approach is Tesco, who have publicly declared their commitment 'to improving welfare standards for all farm livestock'. This claim, which comes from one of the company's own publicity leaflets, is then supported by their proud boast that they are 'the UK's leading supplier of welfare meat products under the RSPCA's "Freedom Food" label'. The same leaflet then goes on to describe the establishing of their own Farmers' Producers Group. Formed in 1996, one of the criteria for membership is that all livestock production systems should undergo an independent audit 'to approved standards of husbandry and welfare.'

Unfortunately, however, their commitment to animal welfare hasn't always been reflected in their willingness to pay the producer an acceptable price for meat produced in this way. A report of a farmers' demonstration outside one Tesco store in December 1998 quoted the following placard, 'Pigmeat - Farmer £47 for one year's work - Tesco £109 for one week's work' (Burrough, D., Jan. 1999:18). This was at a time when the price received by the producer had collapsed to below production costs, whilst the profits of the meat retailers were still increasing (Riley, J. 4th. December 1998:10).

Like Tesco, Sainsbury's have also expressed a commitment to ensuring that all the animal products on their shelves are supplied from farms that uphold high standards of animal welfare. They too took the initiative of establishing their own farm assurance scheme in 1990 under the banner 'Partnership in Livestock'. The aim was to ensure that the 'farmers supplying Sainsbury's are committed to good stockmanship and sound animal welfare... The principles are simple and based on a solid belief in good animal husbandry: all suppliers are required to work to the Sainsbury's Policy on Management of Livestock Welfare and Veterinary Medicine Usage. Welfare standards are verified through independent audit, as well as visits from the processors' and Sainsbury's technical teams. All cattle are reared and managed with care in a humane manner' (Sainsbury's 729/849). With reference again to the crisis in the pig industry in 1998/99, their stance on sourcing pig meat solely from welfare-friendly systems was equally unequivocal. Reporting in December 1998 to the Government's select committee inquiry into the pig crisis they commented, 'All of the pork you buy from our shelves today is already from stall and tether free systems and meat and bone meal free, including the small amount we import from Sweden. On the bacon the same will be true from 1st January and will also include the brands' (Scott, D. Jan. 1999:13).

Such commitment to the sourcing of meat supplies purely from livestock production units where there is evidence of a sensitivity and commitment to animal welfare is to be commended, and in the two cases quoted above appears to be largely adhered to as company policy. In addition, these two examples symbolise the dramatic developments that have taken place in the attitude of most retailers with regard to animal welfare standards. Despite the cynicism of many livestock producers regarding the sincerity of this commitment, the Retail Consortium's agreement of November 1998 does indicate a definite move towards the support of farmers who use such systems of production. Consequently, one hopes that this will eventually lead to even closer co-operation between the producer and the retailer, to the extent that the suspicion and cynicism, which too often currently characterise the relationship, will disappear. Ongoing discussion and debate regarding welfare issues will achieve more for all concerned rather than the control, manipulation and abuse of 'retailer power' that was described earlier.

For such debate to take place, however, there has to be a readiness on the part of both supplier and retailer to understand the trading environment of the other. One reason why the incident quoted earlier, regarding Safeway's imposition of what amounted to a marketing fee, caused anger amongst producers was that the retailer appeared to have little understanding of or sympathy for their suppliers' financial problems. Similarly, retailer demands for supply schedules that bear little relevance to cropping calendars and insistence on product conformity that does not relate to the reality of plants and animals that grow in a variety of shapes and sizes have already caused problems for some producers.

Whilst, therefore, one hopes for closer co-operation between producer and retailer, it has to on the basis of an understanding and sympathetic relationship. If this is not established then one can envisage situations arising whereby, for example, livestock producers will have changes to livestock production systems imposed on them on the basis of apparent yet unsubstantiated welfare improvements that prove to be of no benefit to the stock but at great cost to the producer.

4.6.2   Vegetarianism

Although the apparent growth in the influence of vegetarianism in Western society over recent decades was catalogued in Section 3.6, further consideration now needs to be given both to the scale of this phenomenon and the reasons for it. As individuals forgo the eating of meat for a variety of reasons, it is important that these reasons are identified and that any self-imposed dietary constraints arising out of ethical concerns are especially considered. In fact the validity of such concerns is fundamental to the whole matter of livestock production, for clearly if the vegetarian arguments were to be universally accepted, all animal farming would cease. Similarly, if the ethical arguments against meat eating were proved unassailable then it would be difficult for anyone with a moral conscience either to remain within the livestock industry or to support it in any way.

Whilst some suggest that the number of vegetarians in Britain is growing (see Section 3.6.1), the evidence is that between 1987 and 1996 there has been little change in the amount of meat consumed per person in the UK (see Table 4, page 80). There are slight fluctuations from year to year as one product is favoured over another, but rarely is there a major swing either in favour of increased consumption or less. The exception was in 1995-1996 when the outbreak of Bovine Spongiform Encphalopathy (BSE) in the UK cattle herd was linked to the emergence of new-variant Creutzfeldt Jakob Disease (CJD) in humans. Fears of infection from ingested beef led to a drop in its consumption during the year from 15.3 kg per capita to 12.5 kg (Table 3).

Subsequent fears during 1997-1998 of cross-contamination from cattle to sheep and thus to humans resulted in a similar loss of confidence in the eating of mutton and lamb that year (MLC, 1999:60). The general picture is, however, of continued support for and acceptance of meat as a major part most people's diet in the United Kingdom. As has already been mentioned, this appears to suggest that the supposed growth in the number of vegetarians in Britain over recent years is mistaken. Clough and Kew's description of 'a dramatic upswing' in the number of people adopting a vegetarian diet and life-style for example, would appear from the above to be an exaggeration.

I suggest there is a possible explanation for this apparent contradiction between the claim that more people are adopting a vegetarian diet and statistics which indicate little change in the consumption of meat. It could be argued that as society has become more cosmopolitan so individuals have gained access to a greater choice of food and have consequently adopted a more eclectic diet. Whilst those individuals may be eating more meat, so at the same time they may well be enjoying a wider menu and equally enjoying more vegetarian dishes. Thus, those that would never have considered themselves vegetarian may now be buying more vegetarian meals alongside their purchasing of meat products. It might then be mistakenly assumed from the increase in the sales of vegetarian foods that more people were adopting a vegetarian lifestyle. Indeed, such a conclusion has been drawn from the recent success of food companies specialising in vegetarian products.

The linking of companies producing vegetarian foods with the names of celebrities, such as the late Linda McCartney, has also undoubtedly resulted in a wider public awareness of their products, and of the philosophy that lies behind them. As was indicated in Section 3.6.1 there are indications of growing support for vegetarianism amongst the young, especially amongst girls, and one wonders how far this is influenced or even directed by their interest in celebrity figures. Organisations such as Compassion in World Farming clearly believe that there is some correlation, as they regularly use media stars to great effect in their promotion of the vegetarianism/veganism lifestyle as well as animal welfare. Similarly, Viva, one of the most recently formed vegetarian/vegan organisations, aims its appeal directly at the young by including messages of support from a variety of celebrities representing the worlds of pop, theatre, politics and fashion in their publicity material. Those who endorse the message of their leaflet 'Reasons to go Veggie', for example, include the comedian Victoria Wood, the fashion designer Jeff Banks, and Damon Albarn, the lead singer from the pop group BLUR. As a result one wonders if the effectiveness of such material in influencing the younger generation owes more to the popularity of the communicator than to the actual message itself.

As has already been noted, it seems that the vegetarian diet and lifestyle appeals more to women than men. One suggested reason for this is that vegetarianism has frequently been linked with feminism, whilst meat eating has often been identified with masculinity. Rifkin notes that 'Animal based economies are male-dominated and male-driven whilst plant based economies are far more orientated towards the feminine pole' (Rifkin, 1992:240). Similarly, in 'Meat a Natural Symbol', Fiddel (1991) examines how meat has traditionally been a symbol of dominance within British culture, and suggests that the eating of it can still signify *machisimo,* strength and aggression. 'Killing, cooking and eating other animals' flesh is perhaps the ultimate authentication of human superiority over the rest of nature, with the spilling of their blood a vibrant motif... Meat has long stood for man's proverbial 'muscle' over the natural world' (Fiddel,

1991:65). In contrast, it is suggested that vegetarianism is easily identified with the more feminine qualities of gentleness and pacifism (Fiddel, 1991:110). In support of this the author quotes from the feminist Carol Adams, who remarked that 'being a vegetarian reverberates with feminist feeling' (Fiddel, 1991:199).

Regardless of how valid Fiddel's claim is concerning the symbolism inherent in meat eating, it is easy to recognise the underlying links between these causes of vegetarianism and feminism. As the latter is partially a response to man's exploitation and oppression of women, so the former is often part of a wider moral response to the exploitation and oppression of other species. Animal Rights, which by definition must embrace vegetarianism, sit easily with Women's Rights. As Fiddel comments, 'Vegetarianism tends to be linked with a range of 'progressive' concerns, as an integral part of a personal set of linked beliefs, although the particular concerns will of course vary from person to person' (Fiddel, 1991:200).

Fiddel would seem to identify two strands of thought that cause some women to embrace vegetarianism. In part it is a result of their being able to identify with the symbolism inherent in a non-meat-eating lifestyle, whilst at the same time they are inspired by the moral concerns that have already been discussed in Section 4.2 regarding the possession of rights, and liberation.

The earlier reference to the decline in the consumption of beef and lamb, as a direct result of health scares following the outbreak of BSE in cattle, represents a wider concern felt by many people. Afraid of the food safety implications, many people simply stopped eating beef, and later lamb, until such time as they were assured that it was safe. Whether anyone adopted vegetarianism as a result of the 1996 food scare is unknown, but there is evidence to suggest that the BSE outbreak did lead a number of individuals to make that decision. For example, one of the most controversial figures involved in the debate over the links between BSE and CJD was Richard Lacey, Professor of Microbiology at the University of Leeds. Deeply disturbed by his own perceptions of the health risk posed by BSE in cattle, he consequently reported in 'The Vegetarian' that this fear had caused him to embrace vegetarianism (Lacey, Feb. 1993).

Health fears such as these can be easily exaggerated and manipulated for effect by groups or individuals who want to make a particular point. The following comment, for example, appeared in a paper published by the Vegetarian Society in July 1994. 'If you eat meat, you could easily have eaten something that's been disease ridden, tapeworm infested or tubercular... 95% of food poisoning cases are related to animal products. Meat can contain up to 14 times more pesticides than vegetarian foods. 8 out of 10 oven-ready chickens are infected with salmonella - often caused by feeding dead chickens to live chickens' (Independence, 1994:4).

The publication, still widely available in 1999, encouraged fears that were by then largely groundless. For example, following the BSE outbreak, the inclusion of off-cuts, offal and bonemeal in UK animal feeds was banned in 1996. Similarly, the strict regulations which were laid down in 1995 with regard to the veterinary examination of the carcasses of all animals slaughtered for human consumption, made it almost impossible for diseased meat to enter the food chain. Under the Fresh Meat (Hygiene and Inspection) Regulations 1995 'responsibility for public health and animal health problems in fresh meat (production, slaughter, inspection and marketing) is transferred from Local Authorities to the Ministry of Agriculture and placed under veterinary supervision' (Gracey, 1998:25). Both ante-mortem and post-mortem inspections are carried out on slaughtered stock, and if any trace of disease is found, then the whole carcass is condemned as unfit for consumption.

Despite the assurances that diseased meat cannot enter the food chain, some concerns still exist, however, regarding the levels of pesticide and other chemical residues in a small amount of meat that goes for human consumption. 'The proper use of veterinary medicines, pesticides and other substances should ensure that no harmful residues occur in meat... Regrettably, improper use is resulting in unacceptable residue levels in many products, especially in casualty animals and pigs and poultry' (Gracey, 1998:5). As a consequence further legislation to restrict the use of antibiotics in livestock feed is currently under government consideration. Such a move must surely weaken the case for vegetarianism on the grounds of food safety even further.

In the formulation of the proposed legislation however, it is important that an over-enthusiastic bureaucracy does not ignore the welfare needs of livestock. In reporting on 'the changes proposed in Regulation 13 of the EU Draft Feeding Stuffs Regulations 2000', for example, Bennett (NFU Business March 2000:3) expressed the anxieties of many producers when he suggested that the proposals regarding the administering of vitamins and minerals could have profound negative welfare implications for livestock. The widespread administration of these substances in 'drinking water, as pastes, liquid drenches, sprays or boluses', which is an essential feature of extensive animal husbandry and the current means of treating 'staggers' or magnesium deficiency in dairy cattle, would be banned under the proposed legislation. Bennett suggested that such a move would not only have a detrimental effect on the welfare of stock but would also encourage a move away from extensive husbandry into more intensive systems. For the sake of the stock he argues therefore that these features of the proposed legislation should be resisted.

The recent moves to ensure the safety of meat for eating reflect the legislation that has long been in place with regard to the safety of milk as a food. The Drinking Milk Regulations, which date from 1976, were further strengthened by the Dairy Products (Hygiene) Regulations in 1995. 'These provided for the hygienic production and distribution of milk, the health of the animals from which it is produced, the standard of buildings and water supplies and the standard of animal cleanliness' (Gracey, 1998:24).

Whilst concern over food safety may have been a reason to adopt a vegetarian diet in the past, the legislative measures now in place should reassure consumers that every effort has been made to ensure that food products from animals are as safe as any other dietary ingredients. Unfortunately, some incidents of food poisoning which are attributable to animal products and which can be directly traced to the livestock producer still occur. As with the outbreak of *E.coli* in western Cumbria in January 1999 these incidents are usually as a result of accidental contamination rather than due to the use of a particular production system or method. Such hazards to human health are rare and should not be taken as the rule and as a reason for forgoing animal products.

Of all the reasons for adopting a vegetarian diet, however, there are two concerns in particular that are ethically motivated. The first is the belief that it is morally wrong for humans to kill and digest the flesh of other species. The second is the conviction that a vegetarian diet is not only a more efficient way of feeding everyone but in environmental terms is a more responsible and sensitive way of doing so.

With regard to the former, advocates of the rights view such as Regan (1988) and Linzey (1994) state unequivocally that all animals have a right not to be killed. They then argue that it must follow that we have no equivalent right to deny them that life purely out of a desire to enjoy a meat-based diet. In his criticism of livestock farming Regan (1984:381) condemns the industry for the way in which it 'routinely violates the rights of these animals'.

He then goes on to conclude that 'on the rights view vegetarianism is morally obligatory' and that 'we should not be satisfied with anything less than the total dissolution of commercial animal agriculture as we know it'. Linzey (1994:90) in turn suggests that the avoidance of eating meat and hence the killing of animals is a necessary part of the human quest for a 'higher existence'. It is also a necessary part of accepting the existence of their *theos* rights. 'I...want to conclude that vegetarianism - far from being some kind of optional moral extra or some secondary moral consideration - is in fact an implicitly theological act of the greatest significance. By refusing to kill and eat meat, we are witnessing to a higher order of existence, implicit in the *Logos,* which is struggling to be born in us'.

In response to Linzey's concept of *theos* rights, Leahy (1991:212) suggests that theologically the killing of other species for food,is perfectly natural and acceptable. 'If God created primitive living things like plants for animals to eat (do plants have *theos* rights?), then why should he not have created more complex but still primitive beings which kill each other and may be eaten by man who has been set in dominion over all else?' Webster (1994:15) follows a similar line in arguing that dying and being eaten are necessary and inescapable parts of the natural order to which we are all subject. 'If we object on grounds of conscience to the killing of animals (or nominated species of animals) by man, we do not ensure the preservation of individual lives, we merely change the method of death... if we elect not to kill animals they will still get eaten. It is the inescapable fate of all living animals to be consumed, sooner or later, by something else and used largely for fuel.' He then reminds the reader that even they will suffer the same fate, 'Most humans in history who escaped death at the hands of their fellow humans or other obviously dangerous animals were eventually killed and eaten by micro-organisms'!

Rather than finding this process distasteful and morally objectionable, I find something reassuring in the knowledge that the whole business and design of life and death involves this mutual dependency between the species. It might even be said that there is something sacramental in this dependency, with the consequences of each individual's death leading to renewed life for other parts or members of the biotic community. This is certainly the view adopted by Rolston (1988:93) who as an ecologist has suggested that hunting 'has sacramental value'. 'The unease with which the good hunter inflicts death is an unease not merely with his conscience but with affirming his animality in the midst of his struggles towards humanity and charity, an unease about the dialectic of death with life. The authentic hunter knows suffering as a sacrament of the way the world is made.' Indeed it is this view of life and death which has caused him to ask the question as to whether 'nature [is] at the level of sentient life a passion play?'

Whilst not using religious terminology, Livingstone (1994) appears to come to a similar conclusion when he discusses the inter-relatedness of this community. 'If all of the individual beings in a community share that total, greater consciousness, then it is not unlikely that they may see individuals of their own and of neighbouring species not as 'others' but as simultaneous co-existencies or co-expressions of that place, perhaps as extensions of themselves... They are what they eat and before they eat it... Eater and eaten are equal co-participants.

Each is the community' (Livingstone, 1994:113/114). The death and the subsequent contribution that that death makes to the wider community are valued and respected. Evil exists, not in suffering and death *per se,* as these are both natural processes, but in the attitude that devalues an individual's death and any related suffering and ignores any contribution to the wider good that that death might entail.

Webster (1994:130) applies such thinking to the subject of meat eating when he writes, 'the exploitation of animals, for any reason, can only be justified if we think very hard about what we are doing and how they, the animals, feel about it.' The death/sacrifice of the animal concerned is valued and the contribution that that death makes to the author's well being is recognised and respected. Again to quote Webster (1994:16), 'reverence for life is compatible with a realistic, dignified approach to death' and thus with the moral acceptability of meat eating. As Vidler notes in describing his conversion away from vegetarianism, meat eating 'may be crude; it may be unaesthetic - but it is not wrong' (Linzey and Regan, 1990:197).

If the eating of the flesh of non-human species is not wrong, one must then face the question as to why the eating of human flesh is unacceptable. Cannibalism is an ancient taboo in most societies, and even where it has been practised the reason has usually been cultic or religious reasons rather than dietary (Rolston, 1988:81). The human abhorrence of cannibalism has been due in part to the status that humanity has accorded itself over all other species, or has believed that God has accorded it, and in part to the destructive effect that it would have on culture and society.

The latter point was made by Rolston (1988:81) who suggested that 'humans do not eat other humans because such events interrupt culture; they destroy those superior ways in which humans live in the world. The eating of other humans, even if this were shown to be an event in nature, would be overridden by its cultural destructiveness. Cannibalism destroys interpersonal relations.' As regards the link between human status and the prohibition on the eating of human flesh, one has to look no further than Judaeo-Christian culture, where mankind has been variously described in terms of the 'image of God' (Genesis 1:26), as being made 'a little less than God' (Psalm 8:5) and having all things 'put under his feet' (Psalm 8:6). To eat human flesh would be sacrilegious, even blasphemous. It would be to reduce humanity to the level of all other creatures, and to deny our unique ability to relate to God through each other and the world around us.

Whilst Webster has been able to resolve one aspect of the meat-eater's moral dilemma, the other major issue, which still remains, has been described by Midgley (1983) as the 'most striking reason for not eating meat.' She has argued that 'it is enormously extravagant to use grains, beans, pulses and so forth for animal food, and then eat the animals, rather than letting human beings eat the grains, etc. right away. In the present food shortage, and still more in the sharper ones which threaten us, human interests demand most strongly that this kind of waste should be stopped' (Midgley, 1983:27). Similar concerns have been echoed by others who have suggested that land currently used to graze livestock would be better employed in the production of arable crops. In this way, more people could supposedly be fed. Johnson (1996:59)) has added a further dimension to the debate by drawing attention to the negative environmental impact of some modern livestock production systems, which causes him to question their consequent acceptability.

In the main, traditional husbandry in Britain was based on a mixed farming economy, with arable and livestock enterprises being integrated on the same unit or over a particular local area. Such systems tended to encourage soil fertility and reduce its erosion, a pattern which is now being put into reverse by some of the more specialised livestock systems being employed in various parts of the world. On some beef units in the USA, for example, stocking rates are counted in thousands of animals per hectare of land. Cattle that would have previously grazed naturally on grassland and over-wintered in straw yards are now housed permanently and fed on commercially manufactured feed which is transported directly to them. It is argued that one of the major defects inherent in this dependency on commercially produced animal feeds, is that it is an inefficient and wasteful consumption of food products. As Johnson suggests, the 40% of the world's grain production and 30% of the world's catch of fish used in the manufacture of animal feeds would be better employed in feeding people than livestock (Johnson, 1996:59).

In addition, large specialised livestock units can have a disastrous impact on the environment, especially when it comes to the removal and disposal of animal waste products. The problems are especially acute for those countries such as the UK and the Netherlands, both of which have a densely populated human landscape and a high stocking rate for farm animals. Cattle slurry is notoriously difficult to handle and can cause major storage and spreading problems, whilst large intensive poultry and pig units require large areas of land for the spreading of their waste products. In each case, there are additional problems of social nuisance resulting from noxious smells and slurry seepage into water supplies.

Further anxieties stem from the mineral and chemical residues contained in some animal waste and the environmental damage and health fears that may result from spreading it on fields growing crops. Indeed, Rifkin (1992:4) has considered these problems to be so great that they will only be resolved by dramatically altering the way in which we farm and by changing the sort of food that we eat. 'Dismantling the global cattle complex and eliminating beef from the diet of the human race is an essential task of the coming decades if we are to have any hope of restoring our planet to health and feeding a growing human population. Furthermore, as much of the world's deforestation and desertification can apparently be attributed to the global spread of cattle farming (Rifkin, 1992:24), it would seem that humanity's only hope rests in a universal rejection of livestock production and a widespread adoption of a vegetarian diet.

Arguments such as these would seem to be unanswerable, the case against beef, and by implication all meat eating, completely damning. There are certain factors however, that Rifkin and others who follow a similar line, fail to take into account and which must be considered before a final judgement can be made. He makes no mention, for example, of how beef and dairy enterprises in the UK are largely grass-based grazing units and not permanently housed systems (see Table 7). Whilst there is some dependency on manufactured animal feed in the UK, it is not as dramatic as Rifkin would suggest. No suggestions are offered with regard to possible alternative uses for those large areas of upland Britain that are unsuitable for just about anything other than the production of grass. Such cropping limitations make a mockery of any suggestion that the removal of sheep and cattle would lead to an increase in overall food production. Lastly, there is no mention of the environmental benefits inherent in integrated farming systems, nor of the many intensive farms on which management policy takes account of ethical issues, be they of an environmental, welfare or social nature.

Table 7 **Approximate quantity and value of the major feeds used for animal production in the UK** (Source: Fream 1983:364).

Food source	Quantity (Mt of dry matter)	Value (£m: 1981)
Pasture herbage		
Grazed	20	500
Conserved	12.5	625
Other fodder crops		
Root crops	1.7	150
Kale, etc.	0.6	40
Straw	1.5	35
Cereals		
Home-grown	8.5	850
Imported	3.0	400
By-products	2.0	250
Protein concentrates		
Oilseed residues	1.8	350
Fish and meat meals	0.6	150

This is not a dismissal of the issues raised by individuals such as Rifkin, nor is it a belittling of those concerns which cause people to reject meat eating in favour of vegetarianism for moral reasons. Rather it is an affirmation of an alternative point of view. Livestock production need not be wasteful of global resources, damaging to the environment nor inefficient as a means of growing food. In fact, there is as much moral justification for those systems that adopt an integrated approach to food production, be they intensive or extensive in their operations, as there is for any other. As Webster (1994:139) notes, 'the strongest argument in favour of livestock farming is that based on the capacity of animals to consume food sources that are complementary to, rather than compete with, the needs of man'. Consequently, in those areas dependent on grass as a crop, not only will livestock production make for sensible use of the land but it should also enable the local community to be economically viable.

Furthermore, livestock production need not be an inefficient and wasteful way of producing food. 'When output is expressed as net yield of food for man relative to intake of food in a form that man could eat himself, we discover that the dairy cow can generate 76% more food energy for man than she consumes. This makes her a very valuable animal indeed' (Webster, 1994:139). Webster also argues that efficiency of food production can be achieved in those pig and poultry units which utilise 'the food which we could eat but don't... It is quite possible to operate the scavenging principle on a modern commercial scale, either by processing crop residues to increase their nutritive value for pigs and poultry, or by operating piggeries in association with supermarkets to salvage food that overruns its 'sell-by' date' (Webster, 1994:139). The principle has been well established over many years, and large numbers of pig units have been sustained by the waste from dairies, factories, and packing houses.

With the increased urbanisation of western society, and the subsequent distancing of most people from the production source of their food, there is likely to be a continued growth in the numbers embracing vegetarianism. Whilst some may be motivated by sentimentality or an inability to countenance that animal flesh is acceptable as food, others will change their diet for ethical reasons, some of which have been discussed. The point is that the views of both groups will influence the future of livestock production. It is important, therefore, that livestock producers take account of these feelings and beliefs, that they listen to and respond to the concerns that are raised, and that they present their own case accordingly.

The case for meat eating is defensible, as is the case for breeding and rearing animals for meat. That case is only sustainable, however, as long as the animals concerned are afforded the highest possible standards of welfare that can be achieved. This in turn requires that those responsible for the care of the animals must exercise the highest standards of stockmanship that are attainable.

CHAPTER 5

The Ethical Basis of Livestock Production

The sense of responsibility in farming communities towards the animals they have bred and reared was discussed in Chapter 3. The responsibility covered animal health, their continued welfare and the circumstances of their death, and has generally been contained within the principles of stockmanship, which have been passed on orally from one generation to the next. These principles may well have originated in, and been fostered by, the reverence or worship that our early ancestors showed as hunter-gatherers towards their prey.

As the human domestication of animals proceeded, so the symbiotic nature of the domestication process cultivated an attitude of respect, in that mutual dependency and benefit helped forge a close and caring relationship between man and beast (English, *et al.*, 1992:6; Webster, 1994:4). This view is expressed, for example, by Budiansky (1994:24) who, in suggesting that 'domesticated animals chose us as much as we chose them', recognised such a process of symbiosis. He then concluded that this must 'lead to the broader view of nature that sees humans not as the arrogant despoilers and enslavers of the natural world, but as a part of that natural world, and the custodians of a remarkable evolutionary compact amongst the species'. He suggested that this is 'an example of evolution operating at its highest level - on systems of species, one of which is us' (Budiansky, 1994:162). Budiansky's picture is one in which the farmer engages in a covenant of care with his animals. He looks after them and provides for them and they in turn feed and thus provide for him.

This covenant of care between farmer and animal was originally very basic, limited to simple husbandry such as the provision of food for the stock. There were indications, however, that across a range of cultures man accepted that he had a responsibility to those animals

155

with which he had formed this mutually beneficial relationship. 'In these early village communities we may assume that some men and women developed a degree of reverence for life itself and a degree of compassion - a desire to avoid unnecessary suffering for selected species of other animals' (Webster, 1994:5). In this comparison Rolston (1988:45) sees the emergence of an ethical dimension to the human/animal relationship. You could 'use animals for your needs, but do not cause needless suffering. This ethic is prohibitive on the one side although enjoining care of domestic animals to prevent suffering. It is permissive on the other side; subject to prohibited cruelty, animals' goods may be sacrificed for human interests.'

As the process of domestication proceeded, so too the principle of stewardship emerged as a basis for personal and communal responsibility towards the animals being farmed. Answerable to God, or a number of gods, or the members of his tribe or society, the herder became the subject of an implicit duty to preserve and uphold the welfare of the stock. Something of this is expressed for example in the principle of 'dominion' laid down in Genesis Chapter 1 and expounded in subsequent biblical passages, whereby man as God's steward has to exercise responsibility for and towards other creatures.

The principle of stewardship was examined and expounded by Bailey in 'The Holy Earth'. Originally written in 1915, the work deals with the farmer's stewardship of the land rather than his responsibility towards animals. Much of what Bailey argued relates to the debate regarding humanity's role as steward of the earth and its inhabitants. The farmer is described as 'the agent or the representative of society to guard and to subdue the surface of the earth; and he is the agency of the divinity that made it' (Bailey, 1988:24/25).

Page followed a similar line when she stated that 'a steward has a delegated authority, and Christian stewardship recognises that 'the earth is the Lord's' and not ours to possess or manipulate at will... Stewards of the land are responsible for its conservation, for its lasting improvement and also for the care of our fellow creatures, its non-human inhabitants. No steward worthy of the name would allow exploitation of the creatures under her/his care, nor the exhaustion of vital resources' (Page, 1996:159).

The formal expression of society's sense of responsibility for the welfare of domesticated animals, which was in part passed on to farmers as stewards of the rest of society, was framed within the legislative protection that those animals were accorded in the nineteenth and twentieth centuries and in the founding of various organisations devoted to their protection, such as the RSPCA. Amongst farmers, however, the issue of animal welfare was, and still frequently is, related to the fostering of stockmanship skills, which include the ability to empathise with an animal, to be aware of its needs and conscious of its condition. Such skills are still widely respected within the industry, and the ownership of them remains a matter of pride and pleasure for many concerned with the handling of livestock (Seabrook and Wilkinson, 1998:29). This may well be due, at least in part, to the traditional and cultural context of such skills within communities that are dependent on livestock husbandry.

The value that herdsmen and shepherds place on the skills of stockmanship can be traced back to some of the earliest contemporary accounts of livestock husbandry. Seabrook (1994), for example, in trying to analyse the essential elements of stockmanship, has discovered how old sayings and traditions regarding the care and welfare of animals are still respected today and passed on from one generation to another. Although many farming proverbs originated in the world of superstition and myth rather than that of science, their wisdom is still recognised and their inherent care and compassion for stock are still acknowledged. This is illustrated by Seabrook (1994) 'A sheep's worst enemy is another sheep', 'Shear your sheep in May and you will shear them all away' and 'Lean to the ram, fit to lamb' Although all three proverbs are centuries old, their practical wisdom is still recognised and acted on by shepherds today.

Whilst, however, the long oral tradition evident in such sayings illustrates the history of responsible care for livestock exercised by stockmen through the centuries, they do little to address the current concerns regarding intensive livestock production. Ancient sayings relating to outmoded methods of production and largely passed on from one generation to the next by word of mouth do little to convince the critics of the industry that their concerns are being addressed.

A system of ethical principles for livestock production is needed, which is relevant to current production methods, whilst acknowledging all that is good in the traditional principles of stockmanship and stewardship. In addition, the system needs to establish criteria against which both current and new methods of breeding and raising animals can be assessed and either accepted or rejected. Such is the purpose of this chapter. Moreover, as the principles are established, so too their implications for the livestock industry need to be examined. This process is essential when issues that affect the economic viability of both the individual producer and the industry at large are under discussion.

5.1   Guiding principles

It was argued in Chapter 4 that the confrontational nature of the debate between the advocates of animal rights on the one side and livestock producers and their supporters on the other can often be counterproductive. Too frequently, the protagonists dismiss their opponent's views without appreciating the weight of evidence in support of those views. The animal rights activist, for example, may well be blind to the farmer's genuine care and affection for his stock. Experience has also shown that farmers can be deaf to the validity of some of the activist's criticisms concerning the cruelty and abuse inherent in certain production methods.

Evidence to support concerns regarding the counterproductive nature of confrontationalism in the debate about human/animal relationships has been provided by Serpell (1989). From his examination of the attitudes and feelings of a range of animal users, including farmers, hunters and researchers, he discovered that the criticisms directed at such individuals by animal rights activists were frequently unsound and unhelpful. The response of the rights activists 'to those who exploit animals has been to condemn them outright as insensitive, self-indulgent or cruel. The results of the present study suggest that such labels are neither appropriate nor constructive, and may only serve to increase feelings of hostility and mistrust' (Serpell, 1989:166).

Serpell's conclusions were vividly illustrated by the BBC in the spring of 1999 in the television documentary 'Living with the Enemy'. A committed animal rights activist and vegan was invited to stay with a Welsh hill farming family in an attempt to create greater understanding between the holders of two very different philosophies regarding the human/animal relationship. Despite the apparent efforts of everyone concerned, the conflicts, prejudices and misunderstandings proved to be insurmountable. Neither party seemed to be able to understand or appreciate the other's point of view. No point of contact could be found and no bridges of understanding could be built between the opposing views. As a consequence the sympathetic discussion that could have proved the catalyst for mutual understanding and respect did not take place.

On reflection one wonders whether a neutral environment would have made a difference to the programme's outcome. The farmer was on his home ground, and as a result felt confident and comfortable. The vegan was a visitor to a world that was totally alien to him and as a consequence he exhibited signs of being uncomfortable, uncertain and defensive. This hardly made for an environment conducive to the fostering and developing of sympathy and understanding between the two parties.

5.1.1   The meaning of respect

In contrast to the above, I believe that **the principle of respect** could be the basis for a growing understanding between the two opposing points of view on matters concerning intensive livestock production and that it could provide the required point of contact between the two groups. Furthermore I suggest that it is the main guiding principle for establishing an ethical base for intensive livestock production.

Respect for the animals in human care is accepted as a fundamental principle in the debate on the human/animal relationship by different authors holding very diverse points of view. It is also a basic feature of inter-human relationships, in that sustainable relationships expressed in marriage, parenthood, or nationhood are largely dependent on some degree of respect. According to Broom (1989:81) this is nothing less than as God intended it to be. As a corollary of affection and trust, respect enables family life to function. As an aspect of accepted duty and responsibility, respect underpins community and social life in terms of adherence to social mores and the law. With regard to the human/animal relationship, however, although the wide use of the principle of respect indicates that it is clearly interpreted in different ways by different individuals, the fact remains that it is a point of contact between opposing points of view.

Page (1996:154-156), for example, describes her aspirations regarding the possible relationship between humanity and other species in terms of 'companionship'. She then warns that 'if human beings cannot see themselves as sharing this planet (as companions, literally, share bread) with the rest of creation, then all the features of the ecological crisis, from overpopulation to pollution, will simply spiral on to catastrophe.' The key to companionship, she suggests, is our readiness to see the rest of the natural world 'not as an adversary to be conquered but as an 'other' to be respected' (Page, 1996;155). A. and D. Bruce (1998:132) similarly write about the human/animal relationship in terms of respect, only in their case the focus is firmly placed on the teaching of the Old Testament. This, they suggest, requires that the relationship should be one of 'respect and compassion'. In practical terms this means that 'the taking of life, including animal life, should only be done for serious reasons and in a manner which reflects the moral significance of the animal's life in the prior context of God's created order.'

Clarke (1997:4) also refers to respect as the key to our relationship with other species. His belief is that such respect originates in the human aspiration towards saintliness, which requires that we 'respect the natures of our fellow creatures and the order of which we are all a part.' In a similar vein, Broom (1989:81) refers to Isaiah's vision of the Messianic age and described how the Messiah 'is not seen as bringing exploitation and destruction but respect for all species in their proper place in the world.' This causes him to conclude 'that respect for all living things is necessary'. Following a different line, but still appealing to the same principle, Leahy (1991) concludes his arguments against animal liberation by stating that his support for meat eating, livestock production and animal experimentation is 'perfectly compatible with our treating other creatures humanely and with respect' (Leahy, 1991:253).

It is Regan (1984:233), however, who more fully explores, develops and then establishes the principle of respect as the key to the human relationship with animals. He interprets it in terms of 'a direct duty of justice owed to all those individuals who have inherent value', which under his terms embraces all 'subjects of a life.' These 'have value in their own right, a value that is distinct from, not reducible to, and incommensurate with the values of those experiences which, as receptacles, they have to undergo' (Regan, 1984:236). He then reasons that it is the nature of that value which should provide the criteria for showing respect to other species. At a later stage he defines some of the practical implications of the respect principle, as he understands it. The principle 'requires more than that we not harm some so that optimistic results may be produced for all affected by the outcome; it also imposes the *prima facie* duty to assist those who are the victims of injustice at the hands of others' (Regan, 1984:249).

Whilst all of the above acknowledge the integral part that the principle of respect can play in the discussion concerning our dealings with other species, few define what the principle actually means or entails. Indeed, it is the very vagueness of the term that creates the impression that the many writers who use it might share a common belief and understanding. Reference to the definition of respect provided in the Oxford Dictionary (1959) illustrates, however, that the existence of such a common understanding between these writers is unlikely. Defined as 'to regard with deference; avoid degrading or insulting or injuring or interfering with or interrupting, treat with consideration', I suggest that it is only the last part of the definition that would receive wide assent. Those, for example, who oppose the genetic manipulation of animals, such as Regan (Bruce and Bruce, 1998:128), would be happy with a definition that included 'the avoidance of interference', whilst those, such as the Bruces, who give qualified support to the technology (Bruce and Bruce, 1998:277), would be more likely to omit the phrase. Similarly, judgements as to what constitutes either 'degrading' or 'insulting' treatment of animals are likely to vary considerably. Taking the example of a flock of sheep grazing a field of sugar beet tops, some might well believe that the muddy environment is both insulting and degrading to the animals concerned. On the other hand, as long as the grazing system was well managed, no shepherd would come to that conclusion.

Despite these concerns regarding the lack of clarity in how the term is used I argue that the principle of respect, based on the definition of respect as 'to treat with consideration' provides a degree of agreement between different parties concerned with livestock production issues. Consequently, I suggest that it is an effective starting point for discussion between those parties.

Furthermore, I submit that Rolston's interpretation of the respect principle can be an effective base on which an ethic for intensive livestock production systems can be built. In arguing that 'no ethics is complete until one has an appropriate respect for fauna, flora, landscapes, and ecosystems' (1988:192), Rolston clearly embraces a respect principle. In interpreting what this means in practical terms he proceeds to identify two essential components of respect, a sense of duty and a sense of value. If an object, a creature or a person is treated respectfully then his assumption is that they are valued and that the individual according respect accepts a degree of responsibility towards that which is being respected. As an ecologist he argues that this can even be applied to lifeless, inanimate matter such as dirt in that a person can value it and accept a responsibility towards it in terms of its place within the ecosystem (Rolston, 1988:192).

Although I feel that Rolston's according of respect to things that are lifeless and inanimate is questionable, I believe that his understanding of the respect principle is helpful when applied to our relationship with animals. Whilst, for example, it is difficult to speak of having a sense of responsibility towards a drawing pin or pencil, even if one values them, this is not the case when one speaks in similar terms of animals. In fact I suggest that our valuing of them and our sense of responsibility towards them are well-established features of our relationship with them. I argue, therefore, that Rolston's understanding of respect not only has particular relevance for animal husbandry but that it provides an appropriate base on which an ethic for intensive livestock production can be established.

The two essential features of livestock husbandry are stockmanship and stewardship. My contention is that both are covered by Rolston's definition of respect. **Stockmanship is concerned with valuing an animal as it is, and for what it is.** This is expressed in an empathy with and an understanding of the cattle, sheep, pigs or poultry, under the stock keeper's care. **Agricultural stewardship has been described in terms of a farmer exercising responsibility** towards his/her livestock in all matters concerned with breeding and production as an outcome of the covenant outlined by Budiansky. That responsibility will not only be in terms of a duty to the animals, however, but will also consist of a duty to the consumer in terms of food quality and safety, and a duty to society at large in terms of producing animal products in ways that society considers acceptable.

One aspect of this latter duty may well be showing respect for and adherence to legislation regarding production methods and welfare concerns. In reality, one of the current frustrations of UK livestock producers is the amount of legislation that they have to observe regarding the breeding, rearing and marketing of livestock. Much of it has been formulated recently in response to food safety concerns arising out of the BSE crisis and requirements such as animal movement certificates and Cattle Identification Documents (CID), whilst onerous for the farmer, are an essential aspect of his stewardship to society. I suggest that for the religious believer there is an additional spiritual dimension to stewardship. In the biblical tradition there is a clear indication that humanity has a responsibility under God to act as his steward in caring for his creation and its creatures. For the Christian farmer, therefore, his role as a steward has theological connotations.

My contention is that Rolston's identification of value and duty as the chief elements that constitute respect can be equated with the role of stockmanship and stewardship in livestock husbandry. How one values an animal, for example, can be ascertained from the treatment that the animal receives and its condition under one's care. The outcome is observable and quantifiable, and as a consequence is widely regarded as the yardstick for assessing a stockman's ability and skills. Similarly, stewardship in livestock production is an acknowledgement of *prima facie* duties and responsibilities with regard to every feature of the methods and systems being used. As a result I suggest that Rolston provides the criteria by which the ethical status of livestock production systems can be determined.

5.1.2   What is to be respected?

Whilst the nature of respect has been defined, however, the practical application of the respect principle to the human/animal relationship is still unclear, i.e. what is meant by the phrase 'respect for animals' and what part, feature or characteristic of an animal is to be respected? Clark (1997:4), for example, wrote of 'respect for the natures of our fellow creatures' and Regan (1984:233) of respect for 'inherent value.' Reiss (1998:163) returned to the Aristotelian concept of respect for an animal's *telos*, which 'can roughly be understood as the fulfilled state (or end or goal) of the organism.'

The debate has been further confused by recent developments in biotechnology. Understood 'as all the techniques that use or cause changes in biological material (such as animal or plant cells or cell lines, enzymes, plasmids and viruses), micro-organisms, plants and animals or that cause changes in an organic material by biological means' (de Boer, *et al* 1995:454) biotechnology has raised a number of new concerns. Mankind now has the ability to interfere with life forms in ways that some consider 'unnatural' and 'tantamount to playing God' (Reiss, 1998:161). Arguing that by definition such interference contravenes the principle of respect they judge it therefore to be unacceptable. Others (de Boer *et al.* 1995:455) express their concerns in terms of respect for the integrity of an organism, having defined 'integrity' in terms of 'intrinsic value, autonomy, or naturality.' Whilst, therefore, the genetic manipulation and modification of an animal may leave it essentially as it would have been if there had been no human intervention and may cause it no distress or suffering, the very fact that it is genetically different has caused them to argue that its integrity has been violated (de Boer, *et al* 1995:456).

Whichever word or term is used, be it *telos*, 'inherent value' or 'integrity' the intention appears to be the same, i.e. to establish a definition which fully and adequately describes an animal's state of being, what Russell referred to as 'the animality of animals' (Russell, undated:14). The common aim is the expression of respect for an animal as it is. In many ways 'integrity', as defined above, would appear to fulfil the need, but as has been indicated there are a number of objections to its use. I would suggest, therefore, that Scruton (1994:308) offers a solution when, in referring to Sartre's 'Existentialism and Humanism', he writes of the 'essence' of an animal. Defined in The Oxford Dictionary (1959) as 'all that makes a thing what it is; indispensable quality or element'. I believe that it is the 'essence' of other species which most adequately describes that which we need to respect in our dealings with them.

Applying the definition of 'animality' to the respect principle *vis-à-vis* the relationship between humans and animals, one is immediately freed from some of the anthropomorphic analogies that have afflicted the debate in the past. As Scruton (1996:66) has indicated, the habit of likening animals to humans, rather than increasing their status, actually denies them the respect they are due. He notes that 'to treat these non-personal animals as persons is not to grant them a privilege, nor to raise their chance of contentment. It is to ignore what they are essentially are, and so to fall out of relation with them altogether.' Respecting their 'essence' overcomes such a problem, in that **the animal is respected for what it is as an animal**.

Furthermore, in referring to respect for the essence of an animal, the reference can equally be applied to the individual creature or to the species as a whole. It encompasses the animal's essential characteristics (including its genetic make-up), its needs, its behavioural pattern and its interests. Finally, respecting an animal's essence requires that any questions relating to their welfare be examined as far as possible from their perspective. In responding to Singer's concerns regarding speciesism, Webster (1994:6) wrote: 'I would suggest that he (Singer) is guilty of (benign) speciesism by presuming to first speak for animals without pausing (at considerable length) to ask them what they think and feel.' Respect for a creature's essence, that which makes it what it is, requires us, as far as is possible, to see and feel the world as they experience it. Only in this way can an accurate assessment be made concerning their well-being and welfare and judgements passed on the acceptability or otherwise of production systems on ethical grounds.

As Rolston recognises, the application of this understanding of 'animality' can have far reaching effects. In considering the welfare needs of farm livestock, for example, he appreciates something of their 'essence' when he recognises that although 'food animals are taken out of nature and transformed by culture, they remain uncultured in their sentient life, cultural objects that cannot become cultural subjects.' His conclusion is that 'although tamed, they can have horizons, interests, goods no higher than those of wild subjectivity, natural sentience. They ought to be treated by the homologous, baseline principle, with no more suffering than might have been their lot in the wild, on average, adjusting for their modified capacities to care for themselves' (Rolston, 1988:79). Such views may not be palatable to a number of animal welfarists who would want to make the lot of domestic animals better than that of their wild counterparts.

Objections to sheep grazing a muddy field of sugar beet tops on a wet, cold winters' day, for example, would, under Rolston's criteria and the principle of respecting the sheep's essence, be unacceptable. Whilst the welfarist may consider the circumstances and conditions to be cruel, the sheep are actually in an environment for which they are ideally suited. Their constitution, in conjunction with physical attributes such as a thick, oiled fleece, enables them to cope with the cold and the wet, whilst regular shepherding ensures that they have sufficient food and that their condition doesn't deteriorate. The environment meets the criteria set by Rolston (see above), and the system is therefore deemed acceptable.

Using respect for an animal's essence as a basic principle, whilst understanding essence in terms of all that makes an individual animal what it is, and respect as the expression of value and duty exercised in stockmanship and stewardship, makes for a workable ethic for contemporary livestock production. The implications of applying that principle to intensive livestock systems are examined in Chapter 6.

5.2 Stockmanship and Stewardship – the agricultural expression of respect

If stockmanship and stewardship are the practical components of the respect principle as proposed above, and if they are accepted as the criteria for evaluating the ethical status of production systems and methods, then it follows that the presence of high standards in both areas must satisfy the definition of acceptable husbandry. Obviously, there will still be those who find such a conclusion unacceptable, especially amongst the advocates of animal rights, who as a result of their basic premise find any system of animal farming offensive. Others, however, who do enjoy the pleasure of eating or wearing animal products and who believe that there is nothing morally objectionable in doing so, will find that high standards of stockmanship and stewardship are sensible and acceptable. On the one hand this acknowledges the consumer's appetite for meat, milk or eggs, whilst on the other it expresses their wish that the animals that supply these products should be reared in as humane a way as possible.

To reach this position, however, one must first ascertain what makes for good stockmanship and well-managed stewardship. Once this has been established there will then be economic and welfare implications, for both the farmer and the stock, which will have to be considered. These issues influence the viability of an enterprise. A farmer for example, may be persuaded to introduce higher welfare standards on his unit, but be prevented from doing so by the financial constraints of his bank or the restrictions placed on the unit by the landlord (FAWC, 1998:2).

Another scenario, which is only too common in the present economic climate, is that the standards of welfare which a farmer has already achieved may be adversely affected by the slump in market prices. Poultry farmers, for example, who are currently using battery systems, and who want to change to deep litter or free-range enterprises, may well be prevented from doing so by the low price that their hens or eggs are realising. Similarly, many pig farmers who have prided themselves on their high standards of welfare and stockmanship have been aware of a deterioration in those standards as the current situation in the pig meat market has forced them to cut back on their expenditure on feed, veterinary care and labour. Eighteen months of market prices that have been below the cost of production (NPA 28.1.00:1) have precluded improvements in animal housing, rearing systems, or stocking rates. Realistically the current management issue for the majority of UK pig farmers is survival and the continued process of improving the welfare of the stock has been postponed until there is an improvement in trade.

In addition to establishing the criteria for determining what makes for good stockmanship and well-managed stewardship, other practical realities must also be considered, along with possible solutions. These must include an examination of the support given to the industry and the individual farmer by retailers, consumers and the government. If all three put pressure on the livestock producer to achieve higher welfare standards as an expression of respect for animals, then they too must exercise a degree of responsibility for the economic consequences of these measures ('NFU Business', July 1999:1). Demanding high standards from UK producers, and then importing cheap alternatives from competitors using low-welfare methods, which are banned in this country, is politically inconsistent and morally indefensible (FAWC, 1999:5).

5.2.1 Stockmanship

The Brambell Report identified a number of the skills and personal qualities that are traditionally covered by the comprehensive term 'stockmanship'. These included knowledge of the animal and its needs, an understanding of the husbandry system being used, and the skills necessary to operate that system (Brambell, 1965:56). In a similar vein Ewbank suggests that stockmanship 'involves stock sense (a knowledge of, rapport with and ability to observe animals) and skill in stock tasks (the practical aspects of handling, care and manipulation of animals)' (Ewbank, 1994:8). As regards the individual stockperson this is further developed with the suggestion that they 'should be observant, patient, informed about animals and their needs, skilful in stock tasks, able to recognise health and disease states and be knowledgeable about the workings of environmental control equipment and the measures to take when it fails' (Ewbank, 1994:8).

Having been involved for many years with the rearing of both cattle and sheep, I firmly believe that the quality of stockmanship is crucial to the welfare of the animal. High-quality animal care is dependent on each livestock keeper having the desire as well as the ability to achieve the highest standards of stockmanship of which he or she is capable. This personal observation finds support in the National Farmers Union report on 'The Caring of Livestock' (1995:18): 'We cannot over-emphasise the fundamental importance of first class management and high standards of stockmanship to achieve high welfare conditions on farms, in transit, at markets and at slaughter. No amount of modern and seemingly welfare-friendly equipment and facilities can prevent disastrous welfare consequences in the absence of consistently high standards of management and stockmanship.'

Further evidence for the assertion that high standards of stockmanship are an essential prerequisite for attaining high quality animal care is found in the codes of recommendation for the welfare of farm animals produced by the Farm Animal Welfare Council (FAWC). Regarding the welfare of stock, the prefaces to the Codes state that 'stockmanship is a key factor, because no matter how otherwise acceptable a system may be in principle, without competent diligent stockmanship, the welfare of animals cannot be adequately catered for' (FAWC, 1998:2). This recognition of how crucial the exercising of these skills can be in the management of livestock production systems was expressed earlier in the findings of the Brambell Committee, which enquired into the welfare of animals kept in intensive units. The subsequent report appreciated 'the importance of the standard of management and stockmanship for the welfare of farm livestock' and concluded that intensive systems required even higher standards of care than traditional methods of production (Brambell, 1965:56).

Such descriptions of stockmanship, whilst appreciating the importance of technical knowledge and husbandry skills in the attaining of high standards of animal care, are considered by a number of authors to be incomplete. The reason for this is that as definitions they omit the one quality that many agriculturists concerned with livestock husbandry have considered to be fundamental to everything that stockmanship stands for. This basic component of farm animal care is the ability of the individual stock keeper to form a close and even affectionate working relationship with their animals. English *et al* (1992) describe this in terms of the stockperson expressing such 'sympathetic and caring qualities' that they become a 'mother substitute' for the individual animals (English *et al.* 1992:37).

Indeed, one could argue that this is the concept of animal care that is demonstrated biblically in the vivid portrayal of the selfless devotion of the shepherd for his flock in Psalm 23. Furthermore, this psalm has been held up before many aspiring stockpeople, myself included, as the very epitome of stockmanship. This has been reflected in the comments of a modern-day shepherd who described how 'all who spend long hours in shepherding are committed at lambing time to a degree of care that is always sacrificial of personal convenience. It owes nothing to any calculation of economic gain against time involved. At its most basic that is in many ways the heart of the matter. Good training may make shepherds' efforts more effective, but it will fulfil rather than replace their motivation' (Evans 1996:61).

There is much evidence to support the claim that the ability to form a close relationship with an animal has an important role both in managing farm animals and in establishing high standards of welfare. At the same time, it is clear that there is some disagreement over what causes a person to exhibit this quality in their working relationship with animals. This in turn raises questions as to how a stock keeper can be encouraged or trained in acquiring or enhancing this ability to care.

Seabrook, for example, suggests that stockmanship has a great deal to do with the 'holistic empathy' that the stockperson has with their animals (Seabrook 1996:7) and relates this to the personality of the stock keeper, and his or her inherent ability to care for and value the animal. He describes this empathy as the ability of 'the stockperson to understand the world from the animal's viewpoint and to understand how the animal perceives humans' (Seabrook, 1996:3). This understanding is then communicated in a variety of ways. Touch, for example is an important feature of the relationship, and 'intimate/quasi intimate (level of closeness), spending time touching and patting the animals' is perceived to be an important expression of

170

this empathy (Seabrook, 1996:5). Likewise, from his research into the ways that different stockpeople relate to their animals, Seabrook established that the use of the voice can be a critical factor in the relationship. 'Talking to the animals is a reflection of interaction and empathy. Not only is tone important but so is the pattern of words used and their interaction' (Seabrook, 1996:5). What he considers to be of even greater importance, however, with regard to the stockperson's use of their voice, is that they should talk 'with' rather than 'at' their animals.

This area of Seabrook's research has economic relevance for the livestock producer in that it has illustrated how the use of the voice not only influences the animals' behaviour but also has a marked affect on their productivity. For example, research has shown that the way the person in charge of a herd uses his or her voice effects the amount of milk that the cows produce (Seabrook, 1996:7). This raises the question as to whether there is a correlation between high standards of animal care or stockmanship and increased productivity. This issue is discussed later in this chapter. The attributes of stockmanship are the direct result of the individual herdsperson or shepherd possessing certain personality traits. 'Research...suggests that certain personality traits may be correlated with the type of person who willingly and positively interacts with their animals' (Seabrook 1988:3).

Having established the personality profiles of a number of individuals working with pigs in farrowing units, Ravel *et al.* came to a similar conclusion (Ravel, *et al.* 1996:514). Their research indicated 'the existence of a specific personality profile related to the pig stockperson, who may be characterised as being reserved, serious, unsentimental, not anxious, emotionally stable, conscientious, controlled and introverted' (Ravel *et al* 1996:514). Further study led to the establishment of a strong correlation between these traits and the tasks and responsibilities of the stock keeper and the conclusion that individuals with a specific personality profile were particularly suited to working with livestock.

A long-established and widely held view, still prevalent in many communities dependent on a livestock economy, suggests something similar. It is often asserted that the ability to form a good working relationship with animals is a gift or a personal quality. Comments to the effect that 'stockmen are born not made', that the skills required for looking after stock are 'in the blood', or passed on 'in the mother's milk' are based on the assumption that the required talents for animal care cannot be gained or taught. Indeed, at first glance it would appear that English (1992) supports such a view when he writes: 'This empathy component of the 'art' of stockmanship must be recognised as being very influential... Part of this empathy is in-built at birth by virtue of inheritance' (English *et al.* 1992:40). He then notes, however, that other influences such as experience and upbringing can be just as effective in encouraging the development of this empathy with stock. Indeed, Seabrook comes to the same conclusion. Whilst he acknowledges that the personality profile of an individual stockperson has a bearing on how they relate to their animals he also recognises that appropriate training and monitoring is needed if that care is to be developed and sustained (Seabrook 1984:87).

The conclusions that Seabrook and English arrived at as a result of academic research, others, such as Street (1932), discovered through their experience of practical farming. In the 1920s Street successfully employed and trained a farm pupil from an urban background. Having had no previous background in working with stock the pupil mastered the required skills to the extent that he eventually established his own dairy herd. From this Street concluded that his pupil's success as a herdsman 'was due to his own personality, to his capacity for work, and to his tenacity of purpose' (Street, 1932:169).

In a similar vein, whilst acknowledging that particular personality traits have a role in the making of a good stockperson, Hemsworth *et al.* (1993:33) argue that it is 'the attitudes and beliefs that the stockperson holds about the animals' that exert the greatest influence on how the individual relates to and works with stock. In a later paper, he and Coleman acknowledge the innovative aspects of Seabrook's work in identifying the existence of certain personality traits amongst stockmen and accept that a person's ability to empathise with an animal is an important facet in animal care (1998:7-8). In 'agriculture, it may be the case that there are personality characteristics – for example, degree of empathy or some temperament factors – which predispose people to be good stockspeople…'

However, they deny that this describes the exact nature of or reason for the bond that can be established between an animal and its carer. They are able to appreciate the 'general fondness and friendship that many farmers or stockpeople have for their animals and that 'farmers have long treated and viewed their animals with affection as companions' (Hemsworth *et al.* 1998:20-21). They assert, however, that this is a feature of animal care that is more dependent on a person's beliefs and attitudes than their personality. From a number of studies it is clear that 'observations of stockpeople indicate that the attitudes of stockpeople about interacting with animals are also predictive of the behaviour of the stockpeople towards their animals (Hemsworth *et al.* 1998:74). It is then argued that these beliefs and attitudes can be influenced and changed by specific training to produce better standards of care. Citing the work of Fishbein and Azjen (1974) they suggest that 'behavioural change is ultimately the result of changes in belief' (Hemsworth *et al.* 1998:123). Consequently, 'training stockpeople by targeting for improvement both their key beliefs and behaviours that are influential in regulating human-animal interaction in a commercial setting offers the animal industries opportunities to improve animal performance and welfare' (Hemsworth *et al.* 1998:129).

Despite the differences of opinion between Hemsworth and Seabrook regarding the factors that motivate and thus define stockmanship, it is clear that both share a basic understanding as to what it means in practice. Essentially, it is about a person's ability to value and thus care for the animals in their charge. As has already been indicated, Rolston argues that such valuing is one of the two basic components in the principle of respect. One values and cares for that for which one has respect. I therefore suggest that the stockperson's knowledge and understanding of an animal's nature and needs, and the ability to empathise with that animal and to care for it, are indications of a respect for what has been defined earlier as the essence of that animal.

Furthermore, English *et al.* (1992) suggest that good stockmanship requires and facilitates a reciprocal respect from the animals themselves. 'A key feature of good handling is that the animals and humans must have respect. Insufficient respect by the human may result in the animal becoming more frightened of humans and sometimes attacking in defensive threat. Insufficient respect of the human by the animal will result in the human being completely manipulated by the animal and, as a result, the animal may become aggressive when it does not get its own way' (English, *et al.* 1992:32). As a result, the design of efficient, and therefore by definition productive livestock management systems must focus on those features which enable this mutual respect to be established. Animals and stock keepers must be placed in an environment that enables them to form a relationship in which the one values the other.

In traditional extensive husbandry systems, interactions between the animal and its keeper would seem to have been an easy task. Overall, the number of animals for which one individual was responsible was smaller than in modern intensive production units. The size of a flock or herd was limited by the physical constraints of the stockperson's having to feed them, tend them or milk them by hand. No matter what size the unit might be, the same arithmetic applied. More cows or more sheep only required more hands to look after them. The ratio of humans to animals remained the same. The consequence of this was that each stockperson was better placed than their modern counterpart to get to know the individual animals for which they were responsible.

Since the Agricultural Revolution of the last century, however, all this has steadily changed. As the hand-milker gave way to machine and free-range poultry units were replaced with battery systems, so the required ratio of humans to animals for managing livestock also changed. Indeed, the change continues, with units continuing to increase in size, whilst the number of people required to operate them continues to decline. In 1804 when Arthur Young reported on the state of agriculture in Hertfordshire he noted that the size of pig herds ranged from 25 pigs to 200 (Young 1971:212). In 1989 the annual records of the Ministry of Agriculture Fisheries and Food reported that the average UK pig unit carried 355 fatteners and 58 breeders (MAFF 1990:11). By 1997 this had increased to 489 fatteners and 81 breeders (MAFF 1998:12). A similar picture emerges regarding the size of dairy units, with 12 cows being the average size of a milking herd in

1804 (Young, 1971:183), 62 cows the average in 1989 (MAFF 1990:11) and 66 cows in 1997 (MAFF 1998:12).

As herd size increased, so the number of agricultural workers steadily declined. In 1986-88 the total agricultural workforce in the United Kingdom (excluding Scotland) amounted to 674,000 individuals. By 1997 this had declined by 70,000 (MAFF 1998:14). The overall result has been a dramatic increase in the number of animals for which an individual stockperson is responsible. This in turn has meant that the stockperson is unable to spend as much time with individual animals as would have been possible in the past under previous regimes.

Stock keepers of a previous generation would have been managing production systems which would frequently carry animals to a greater age than those of today. This would enable the herdsman or shepherd to become emotionally attached to his stock in a way that would be difficult in current production regimes. In the years leading up to the Second World War, beef cattle for example would have been fattened for up to three years to produce a heavy carcase that would be considered too fat for many of today's consumers (Donaldson 1969:184). Current practice requires the beast to be slaughtered at half that age, when the meat is lean and tender.

Likewise, the current preference for lamb rather than mutton is also comparatively recent. As a result, the older mutton-producing systems, which would fatten a lamb for up to two and half years, would enable the shepherd of the past to know each of his sheep. In contrast, whilst a modern shepherd may know the identity of individuals in their breeding flock, the chances of them getting to know the lambs being fattened for killing at three or four months is remote. Similarly, modern high-yielding dairy cows may well be culled from the herd after producing four or five calves. This may be due to the farm adopting a rigid seasonal calving policy, or simply because the strain of high milk production has exhausted the animal. In contrast, a more liberal calving calendar and a lower milk yield in the past meant that dairy units might well carry cows that were twice that age. Indeed, I can remember milking a cow in the 1960s that was reputed to be twenty years old. Over those two decades, as that animal was largely milked and fed by the same person, and as that same individual probably calved her many times over, it is not surprising that an affectionate bond was established. So much so that when the cow eventually had to be culled he broke down and wept.

It is tempting to quote a story such as this in support of more traditional methods of farming, arguing that such an affectionate relationship between human and beast must make for greater respect between the two. A herdsman who cared enough to weep over the passing of an individual animal clearly cared very deeply about it and valued it. In some ways this was obviously true, but as regards respecting the 'essence' of the animal this might not always be the case. Close affectionate relationships, such as the one just quoted, could well lead to the herdsman adopting an anthropomorphic attitude towards the animal rather than appreciating its true nature and needs. This is evidenced in the way that cows in small herds were frequently given names and credited with having almost human attributes. Older, well-established members of a herd would become pets and treated as such. The same could often be said for pigs that were kept in small units, especially those that were reared in the sties that were often found next to the cottages of rural workers, or hens that were kept in the small neat runs that often adorned a cottage garden.

Rearing regimes such as these, based on close relationships, suggest the presence of high standards of care and therefore welfare. The fact that the true nature and needs of that animal were often ignored or not fully comprehended meant, however, that the opposite would often be true. My own childhood memories of small, poorly ventilated, stygian dark byres containing three or four cows permanently tethered to their stalls throughout a long winter are reminders of an unthinking regard for the animals' needs. Similar recollections of a fattened porker confined to a tiny and often filthy sty, and of a few hens scratching in the dirt of a tiny immovable run, adds further weight to the contention that affectionate anthropomorphism need not by definition be equated with good stockmanship.

Rather, I suggest that a well-trained and highly motivated herdsperson, working in a well-designed intensive system, is likely to provide their animals with a much higher standard of care than their predecessor. Equipped with a firm grasp of an animal's behavioural, nutritional and physical requirements they work in an environment specially designed to take full account of what they and the animal need. In fact, these two latter concerns are now paramount principles in the designing of livestock housing.

When Leaver described the environment of the modern dairy unit he commented that 'the requirements of a housing system include those necessary for the well being of the cows and those needed to provide an economic workable system for management' (Leaver 1994:32). Similarly Sainsbury noted that the 'design of the animal accommodation at every point should give full attention to the stockman's needs. The building should be well lit for inspection purposes, with natural lighting used wherever possible' (Sainsbury 1986:12).

Sainsbury expresses equal concern for the welfare of the stock, and reference is made to a wide range of housing designs that take full account of the animal's needs in terms of disease prevention, cleanliness and comfort. The key to all the designs that he advocates would appear to be his appreciation of the nature and thus the essence of each animal. This in turn ensures that the criteria for assessing the welfare of each individual are understood from the animal's perspective rather than that of its human keeper. The result is that a great deal of thought and understanding of an animal's needs are likely to have been applied in the designing of many modern livestock units. One cannot help but compare this with the more traditional byres and sties described above and in the process draw the conclusion that many modern units will be more in keeping with the principles of stockmanship than those of the past.

Whilst the skills and knowledge of the modern herdsperson have undoubtedly improved standards of animal care when compared to those of his/her predecessors, the issue of herd or flock size, which was raised earlier, still needs to be examined. As has been indicated, recent decades have witnessed a tremendous increase in the number of animals for which one individual will have responsibility. The stock keeper of the past, caring for a small number of animals was able to identify each one and to know them as individuals. This would seem to facilitate the exercising of stockmanship skills in a way that at first sight appears to be impossible in today's much larger and more impersonal production systems.

Having reflected on the possible bond that could develop between the animal and its keeper in the smaller extensive husbandry systems of the past, English *et al.* (1992) noted for example that modern intensive units created particular difficulties for the stockpeople working in them. 'It is possible that, as the size of herds and flocks under the one stockperson has increased over recent times resulting in the blurring of identity of individual animals, and increased pressure on the attendant, he may have less time to talk to his animals and give them 'praise with the voice' as appropriate. Such conversing of the stockperson with his animals... undoubtedly plays an important part in developing mutual respect and empathy and in increasing ease of handling' (English *et al.* 1992:34).

This poses the question as to whether modern intensive systems, in which such intimacy is impossible, must by definition exclude the practice of good stockmanship. If this is found to be the case, and if as has been argued stockmanship is accepted as an essential expression of that valuing of a farm animal which is required by the principle of respect, then it must follow that such systems are ethically unacceptable. Indeed, the thrust of many arguments against the methods of livestock production used in modern agriculture is that their very size and structure makes it impossible for the stockperson to value the individual animals in any meaningful sense. As Ewbank commented, 'one of the criticisms of modern intensive systems is that far too many animals are being looked after by far too few stockmen' (Ewbank, 1994:8). Whilst there are some systems that may well be criticised on these grounds, I would argue that responsibility for a large herd of flock does not of necessity prevent a stock keeper from expressing high standards of stockmanship.

I believe that I am able to care about other humans without having to know them as individuals in any personal sense. Even when confronted by a crowd of people, I believe that I am still able to value them, appreciate them, and therefore to a certain extent to care about their needs as individuals. As long as I know of them and know something about them in terms of their interests, I believe that I am in a position to understand, respect and value who and what they are. For example, the manager of a large manufacturing plant will not know the identity of each of his employees, nor is he likely to have a close personal relationship with more than a few of them. This does not prevent him from valuing them, nor does it stop him from either appreciating their worth or from caring about their needs. Rather, as a caring employer he takes an interest in their welfare, he acknowledges his responsibility towards them, and as they all share a common interest in ensuring the firm's profitability he feels he can identify with them.

In a similar way I suggest that, whilst the milker responsible for a herd numbered in hundreds will never know each individual cow in the same way as the herdsman mentioned earlier, he or she is still able to value the animals and thus care for them. Their sense of respect for the individual animal is not based on numbers, but on the principles that direct the way in which they as a herdsperson relate to their stock. Indeed they may value and care about each cow without ever giving that animal a name and therefore an individual identity. Instead, they will express that care in the way that they handle the cow, speak to it and minister to its needs in the milking parlour and out in the yard or field. Research has indicated that this close interaction is then likely to result in increased milk yields, less stress in the dairy herd and improved standards of health in the individual animals (Seabrook 1984:87).

Further research conducted by Seabrook *et al.* (1995), regarding the individuality of pigs in large intensive systems, discovered that even in these seemingly impersonal units similar relationships between animal and herdsperson could be established. The pigs were not only able to identify their different stock keepers but were also able to form relationships with them, recognising that 'the stockperson plays a significant role in the life of the animal, providing food and stimulation' (Seabrook *et al.* 1995:47).

Hemsworth and Coleman (1998) take this further in suggesting that, rather than creating an environment in which the stock keeper is unable to identify with their stock, many modern systems actually facilitate the establishing of a close relationship between man and animal. Their reasoning is that such systems require the stockperson 'to regularly monitor animals and their conditions and impose routine husbandry procedure' to an extent that is unknown in extensive stock keeping regimes (Hemsworth and Coleman, 1998:39). To support their argument they compare the shepherding of a wide-ranging flock of hill sheep with the maintenance of an intensive meat-chicken unit. Under the former regime, there is little contact between sheep and manager, whilst under the latter the stock are inspected up to six times a day. Their conclusion is that 'even though the number of animals managed by each stockperson has substantially increased in modern production systems, farm animals in these systems receive frequent and, at times, close human contact' (Hemsworth, and Coleman, 1998:40). Rather than intensive systems militating against the exercising of good stockmanship, they argue that such systems will frequently encourage the process. 'In intensive livestock production, there is frequent and often close contact between stockpeople and animals, particularly young animals and breeding animals, and as a consequence of these interactions, many of which are far from superficial, long-term relationships develop between humans and animals' (Hemsworth, and Coleman, 1998:61).

It has to be recognised, however, that there are certain current livestock production systems that make the establishment of such relationships difficult. No matter how frequently the needs and environment of hens kept in vast deep litter systems or housed in battery units are monitored, for example, it would still be difficult for a stock keeper to identify with the needs of individual birds. These concerns will be addressed in the next chapter. The argument remains, however, that the size of modern livestock units does not necessarily preclude the expression of stockmanship and the valuing of the individual animal.

The second reason for dismissing the suggestion that the exercising of stockmanship qualities is impossible in modern production systems is that, contrary to common assumptions, many current systems have actually encouraged positive developments in standards of animal health and care. As regards the health of animals reared under these regimes, any comparison with their predecessors of even a few decades ago will show distinct improvements. Again, there are exceptions, and such methods of production will be deemed on this basis to be ethically questionable, and as a consequence will be discussed in the next chapter.

Amongst the more recent innovations in livestock production methods, the introduction of Automated Milking Systems (AMS) stands out as a technological development which has particular implications for standards of animal care and especially for the expressing of stockmanship skills. Indeed, Webster (1994:176) suggests that 'it may prove to be one of the most successful marriages between high technology and animal welfare.' For those accustomed to the well-established assumption that a milker can only know their cows by handling them in the parlour twice a day, the notion of leaving this to an automated system appears to be a step in the wrong direction. Added to this there are likely to be concerns amongst herdspeople regarding a perceived loss of purpose and job satisfaction. From their interviews with a sample of 261 employees working on dairy farms, Seabrook and Wilkinson (1998:29) discovered that 61% rated milking as their 'best/most enjoyable task.' They also noted that it was considered to be 'both important and enjoyable'. The correlation between the point score for importance and the number of people citing milking as the best liked task was +0.79. It would seem likely, therefore, that the introduction of AMS onto UK dairy farms is likely to meet some resistance from those currently milking cows.

In contrast, the advantages of AMS in terms of animal welfare are immediately obvious, in that automated systems enable the individual cows to choose when they want to be milked and fed. Rather than being forced to adhere to a human imposed milking regime of two or three times a day, they have round the clock freedom to enter the unit when they want to. The subsequent welfare implications were realised in the research conducted by Prescott *et al.* (1997), which concluded that AMS provides farmers with the opportunity to engineer choices into an animal's environment (Prescott *et al.* 1997:80). Indeed Carter and Stansfield (1994:48) comment on 'the considerable benefits on farms where the owner or stockperson could use their time to greater advantage in management and health checking than in the routine chore of milking'.

This potential for the development of and improvement in the quality of stockmanship exercised by the herdsperson using AMS was also appreciated in a report on the installation of such a system at the Agricultural Development and Advisory Service's (ADAS) Bridgets Research Centre. Initial results indicated that the system not only relieved the herdsman of the monotony of milking but that 'any time saved is taken up observing and monitoring the cows' ('Farmers Weekly' 1st. Aug. 1997). Rather than having to interact with their stock in the restricted environment of the dairy parlour, with the incessant demands that this places on the herdsperson, in theory they now have more time to move freely amongst the herd. This should enable them to identify the characteristics of individual animals, to observe particular behaviour patterns and to notice emerging health or welfare problems.

A report on the increased use of such systems on dairy farms in Holland ('Farmers Weekly' 10th. Nov. 1998) made reference to additional welfare benefits that were being recognised. Comments were made to the effect that the cows were 'more at ease', 'quieter' and apparently more 'comfortable'. Further welfare benefits were appreciated in terms of human welfare in that AMS improved the working environment for the farmer and his family in that automated systems provided them with greater flexibility regarding how they spent their time. Family life was no longer dominated and restricted by the demands of routine milking.

In their concluding comments on the potential benefits of robotic milking systems, Carter and Stansfield (1994:54) make the point that such systems require the employment of 'well-trained and motivated stockpeople'. Other commentators have come to the same conclusion, not only with regard to highly sophisticated technological developments such as AMS, but with regard to all intensive systems. Whilst Sainsbury (1986:4) suggests 'that able stockmanship is in measure art rather than science' he still concludes that 'modern intensive husbandry is full of complex equipment and methods, and training is essential for its proper application.'

In a similar vein the report published by the Brambell Committee not only recognised that intensive livestock production required a highly trained work force, but then recommended that as a result 'the provision and scale of appropriate training facilities should be reviewed' (Brambell, 1965:56, 63). Others suggest that post-war developments in agricultural education have focused on the provision of technical knowledge rather than the more perceptual skills associated with stockmanship. This is noted by Hemsworth and Coleman (1998:14), who comment that over recent decades agricultural training has tended to place the emphasis on 'technological innovation, especially in areas such as housing, nutrition, breeding and health' with little provision for training in 'stockperson skills.'

This failure within agricultural education has partly been due a widely held belief that the skills defined by the term stockmanship are related to outdated ideas and practices, to the farming of the past. As a result, Seabrook (1988:2) argues that 'stockmanship skills, including animal and human interaction, must be projected with progressive, efficient and positive images. This training must at all times put an emphasis on improving perceptual skills.' I believe that this recommendation needs to be acted on by the industry at large, and by its academic institutions in particular.

Furthermore, I believe that, related to the demand for training in perceptual skills, there also needs to be a parallel demand for the provision of training in the principles that undergird them. Stockmanship is one of the two components of the respect for animals that is the ethical basis of livestock production. Therefore, to make the ethical basis practical reality, stockmanship must be developed to a higher skill level than it often is at present. As a consequence, a knowledge and understanding of the nature of those animals must be an integral part of the educational process which underpins the development of the stockperson's skills. Only then will the needs of the stock be fully appreciated and they in turn be truly valued.

5.2.2   Stewardship

Whilst the exercising of high standards of stockmanship demonstrates a caring attitude towards livestock and indicates that stock keepers value the animals in their charge, the principle of respect demands more. If the valuing of an animal and the subsequent level of the care it receives are solely dependent on an individual's ability to empathise with and feel affection for that animal then the requirements of the principle of respect are not being fully met. Human affections are notoriously fickle and are susceptible to such a wide range of external influences that they are bound to prove a most unstable foundation for any ethical system that is built upon them. Any attempt, therefore, to use empathy and affection in isolation as an ethical base for either intensive or extensive livestock production is bound to prove frustrating. One might value the animals one day and on the next find that because of business anxieties or family problems one hates the sight of them. If the quality of care that the animal receives is purely dependent on one's ability to empathise with them, then on that particular day the animal's welfare is likely to suffer.

In reality, the care most stock keepers show for the well-being of their animals transcends any personal feelings or circumstances. They have a sense of responsibility for their stock which, whilst it may involve affection, is also to do with how they value those creatures for what they are. I believe that because a good stockperson understands the essence of their animals, as it has been defined in Section 5.1.2, so they are able to appreciate and value the characteristics and qualities that constitute a sheep, a cow, a pig or a hen. As a sheep breeder I value the mountain breeds such as the Swaledale or Scotch Blackface for the way they are adapted to life in a hostile environment and for their ability to survive under the harshest conditions. But more than that, I appreciate and admire the qualities inherent in those sheep as a species - their care for their young and their ability to exist in some of the most inhospitable terrains. Similarly, having worked with cattle, I value the Holstein-Friesian, but not just because it is a prolific milker but because of the bovine qualities it exhibits such as patience, inquisitiveness and trust.

Stockpersons value an animal by appreciating it for what it is and by realising the place that only it has in the overall system of things. They recognise and value the contribution that it makes as a sheep or cow to the life of the countryside or to society at large. They are sensitive to the fact that the landscape and the environment would be diminished without those animals, and as a result of all these factors they have a deep sense of responsibility for their care and well-being. Valuing the animals leads to an acceptance of responsibility and to a sense of duty for their care that demands that the welfare of the animal frequently takes precedence over the comfort or safety of the stockperson. In many cases the responsibility is exercised up to the moment of slaughter, and Webster (1994:16) notes how one farmer travels with them to the abattoir and then walks 'with them up the race to the stunning pen so that they remain calm and do not suffer right up to the point of non-existence.'

This understanding of a stockperson's responsibilities is epitomised in the biblical picture of the good shepherd, 'who lays down his life for the sheep' (John 10:11). This total dedication of a shepherd to his sheep and the sense of duty that it implies is described by Evans (1956:52) when he writes, 'Shepherding is a hard lonely life, a twenty four hours a day job when a man can afford to have nothing in his 'hid' but sheep. Yet, in spite of this, something remains about a shepherd's life that appeals: perhaps it is that his life is more selfless than most.' The shepherd values his flock and each member of it and consequently feels a deep sense of responsibility for their care.

Rolston (1988:230) recognised this link between value and duty when he stated that the very valuing of anything or anyone involves the acceptance of responsibility towards that person or thing. To speak of valuing someone and then to show no interest in their welfare or circumstances would be hypocritical, a denial of the declaration of value. Indeed, a number of charities use this as the basis for challenging the public to support their work. If we value children, or the elderly, or animals, how can we see them suffering and do nothing about it? Rolston's concern related to the valuing of our environment, and he argued that this imposed a consequential accepting of responsibility for its protection on the valuer. I suggest that a similar case can be made regarding our relationship with livestock. In valuing one's animals and expressing that valuing through the exercise of high standards of stockmanship I believe one is implicitly assuming a duty of responsibility for their welfare.

In traditional agricultural terms, such a sense of responsibility towards one's stock is frequently defined by using the concept of stewardship, as was suggested earlier in this chapter. When applied to the care of livestock, agricultural stewardship establishes the various demands of duty and responsibility to which a stock-keeper is subject. Furthermore, stewardship stands alongside stockmanship as one of the two components that constitute my proposed ethical basis of intensive livestock production. A principle of respect requires that the one expressing respect should both value that which they respect whilst also accepting a responsibility for its care and protection.

Moreover, I suggest that **the concept of stewardship requires that the exercise of this responsibility constitutes a solemn duty**. Indeed, this would seem to be the conclusion of the authors of the Church of Scotland's report, 'While the Earth Endures' in which it is stated that 'a Steward does not ruthlessly exploit or exhaust the resources entrusted to him. He does not shelve or try to shift his responsibility for any situation, nor let it grow worse through neglect. He thinks not solely in terms of the present, but has a responsibility for the future' (cited in ACORA 1990:14). The practical implications of regarding stewardship as a solemn duty are recognised by Midgley who states that 'the whole point of a duty is that it is bounden or binding. It may be slight; it may be overridden by other duties, but so far as it goes it must surely bind us' (Midgley 1983:52). If stewardship imposes a duty on livestock keepers to care for the well being of their stock, then I submit that it is binding. The duty of care is not an option but a prerequisite of responsible livestock husbandry.

The imposition of this duty of care on the stock keeper has, I believe, three separate sources - to God, to society and to the animals. Each has to be acknowledged as an integral part of the total picture. Whilst some stock keepers recognise a duty under all three, others may believe that their duty is to but one. Ultimately, however, the implications for animal husbandry will be the same, in that the demands of all three require the exercising of high quality animal care. Furthermore, as it has been recognised, the duty to observe these standards is binding.

Whilst the Church has been accused of encouraging the exploitation and suffering of animals (Singer 1995:192), Thomas (1984:153-4) argues that it was in fact the widespread influence of Christian teaching that first inspired the cause of animal welfare in England. He describes how 'one single, coherent and remarkably constant attitude underlay the great bulk of the preaching and pamphleteering against animal cruelty between the fifteenth and nineteenth centuries.' Over the course of four centuries Puritans, Quakers, Dissenters, Latitudinarians, Methodists, all preached that God has given man a solemn duty under which he 'was fully entitled to domesticate animals and to kill them for food and clothing. But he was not to tyrannize or to cause unnecessary suffering. Domestic animals should be allowed food and rest and their death should be as painless as possible' (Thomas, 1984:153).

The source of this attitude is traced back to the Old Testament doctrine of man's stewardship over creation, to the belief that man is answerable to God for the way in which he uses or abuses that which has been divinely entrusted to him. The reference to mankind being given dominance over Creation in Genesis Chapter 1:28 is interpreted in terms of his being 'seen as a kind of managing bailiff for God. An agent with authority to use available resources to support all those dependent on the estate, but bound to give account of his use and custody of what is not his own' (Jones, 1988:18). This sense of divinely initiated duty towards one's stock was to have obvious practical implications for those livestock farmers who professed a biblical faith. 'The good husbandman... would ensure that his animals were well fed and watered and that they were allowed to rest on Sunday. He would also be careful not to lose his temper with them or strike them in anger if they failed to perform according to his expectations... Considerate treatment of animals became a religious obligation' (Thomas, 1984:154).

I contend that over the course of centuries this application of the biblical principle of stewardship to livestock husbandry has become an integral feature of UK agriculture. I would argue that in those families that have been concerned with livestock production over the course of several generations it has become culturally ingrained (Rollin 1993:8). Indeed, experience has shown that many farmers from such families, whilst they make no profession of religious faith, still identify with this role of steward and view the tasks of shepherding their stock as a sacred duty. They would rather put themselves at risk than deny that duty. Indeed, in the current agricultural climate many have done so financially, as they have continued to feed and medicate their stock even when they know that the financial returns from that animal do not warrant such expenditure.

For the stockperson who is a committed Christian believer, that sense of duty under God becomes paramount. As Jones (1988:18) comments, 'this is an awesome responsibility that must not be shirked, nor must it be abused... Man is to reflect God's creativity and character within creation'. Having stated that 'Christian stewardship recognises that 'the earth is the Lord's' and not ours to possess or manipulate at will', Page (1996:159) similarly concludes that 'no steward worthy of the name would allow exploitation of the creatures under his/her care.' Likewise, the authors of the Church of Scotland report 'While the Earth Endures' (1986:10) recognised that the believer who takes the role of steward seriously must reject certain attitudes that conflict with that role. This must include 'anything which could be deemed exploitation' and 'making present profitability the sole effective criterion'.

None of these commentators identifies what they mean by either abuse or exploitation, and the definitions of both are left to the individual's personal judgement. In terms of practical husbandry, therefore, they are of little help in establishing criteria for the stockperson to follow regarding what constitutes appropriate and acceptable methods or system of production. Whilst some, for example, will consider certain practices ethically admissible in that they conform to their understanding of what constitutes their duty to animals, others may well consider them to be examples of exploitation and/or abuse. This is one of the issues that will be examined in Chapter 6, when some current husbandry practices are discussed and suggestions made regarding their acceptance or rejection on the basis of the criteria established under the principle of respect.

In 1990 the Church of England published their report 'Faith in the Countryside' (1990) and for its definition of stewardship the authors cited the definition previously arrived at by the Church of Scotland (see above). They then moved the discussion further on, however, by suggesting that stewardship must be exercised not only in terms of our relationship with God, but also in terms of our wider relationships with each other. Community is identified as one of the core Christian values, and it is contended that as we have a duty to God so we have a duty to each other. 'Our identity is defined relationally, and there is also a corporate responsibility for justice, fairness, peace and right relationships within which we contribute our voice... It is part of our response to God to help to create conditions where others in community experience care and well-being' (ACORA, 1990:22).

From this, it is possible to identify two further impositions of duty on the livestock producer regarding the care of that stock. As we live in community so farmers who rear animals for the use and consumption of that community place themselves in a position of responsibility under that community. Effectively, they are rearing the animals on behalf of the community that will consume the meat. As a result those farmers are society's stewards. Society has delegated responsibility to them for the care and well being of those animals.

Furthermore, domesticated animals have been have been brought into the human community. Budiansky (1994 :14) suggested that the process of domestication was not imposed on them by mankind but was a feature of evolutionary community development, part of 'a mutual strategy for survival'. As man fed and looked after those animals that sought the protection that the human community provided, so those same animals provided that community with food, clothing and transport. This mutually beneficial arrangement was effectively a covenant relationship in which humans accepted a responsibility to feed and protect the animals that supplied them with some of their most basic needs. These relationships constitute the other two areas in which livestock producers have the duties of stewardship imposed upon them – responsibility to society and to the animals themselves.

If, as has been suggested, livestock producers are society's stewards and that as a consequence they have a duty of responsibility for the care of animals delegated to them by the rest of society, then society must equally have a reciprocal duty to monitor and direct those standards of care. Delegation requires responsible management, and with regard to British agriculture, this has long been the role of animal

189

welfare groups, government, and the legal system (Webster 1994:259-261). The development of animal welfare legislation in the United Kingdom was traced in Chapter 2, where I indicated how the concerns of society for the well being of animals were given voice by organisation such as the RSPCA which, through their representation to government, brought about the framing of animal welfare legislation. This in turn led to those animal husbandry practices and systems which society found unacceptable being outlawed. This process still continues, as is illustrated by the path onto the statute books of recent legislation regarding farm animal welfare. This legislation includes the banning of sow stalls in pig production systems under the Welfare of Livestock Regulations 1994 (SI1994 No.2126) and the phasing out of battery cages under the European Union's Laying Hen Welfare Directive of 15th. June 1999 (EU Directive 99/74/EC).

A major contributor to this process, and one that is often unknown outside of agricultural or governmental circles, is the Farm Animal Welfare Council (FAWC). The Council's origin as a practical and significant outcome of the Brambell Committee's recommendations has already been noted in Chapter 3, but it is important at this stage to appreciate the role that it continues to exercise in terms of monitoring livestock production methods on behalf of the wider community.

Established by the Government in 1979, its terms of reference 'are to keep under review the welfare of farm animals on agricultural land, at market, in transit and at place of slaughter; and to advise the Minister of Agriculture, Fisheries and Food and the Secretaries of State for Scotland and Wales of any legislative or other changes that may be necessary' (FAWC, 1999:2).

Over the years FAWC has published welfare codes for each sector of the livestock industry, including the farming of more exotic species such as deer and ostrich. These have set the standards to which livestock producers are expected to adhere regarding the care of their stock, and are now widely used in agricultural education as a means of helping potential stock keepers to appreciate the responsibilities that society places on them. In addition, each producer with employed staff has a further responsibility to ensure that each member of staff is aware of their duties to the stock as required by the relevant code.

The well-established and widely accepted credibility of the welfare codes have been established as a result of the painstaking research and consultation process which lies at the heart of FAWC's activities. When a welfare issue is raised or when the government seeks guidance on a particular matter, a wide range of interest groups is consulted, scientific evidence is commissioned and examined, and the advice of experienced agriculturists is sought. This process may then lead to the publication of reports, the framing of new legislation or the presentation of proposals to government for further areas of research regarding the welfare of farm animals.

This process indicates, however, that just as the farmer has a duty towards the wider community regarding the care of animals, so too, through FAWC, society accepts something of its reciprocal duty to monitor the standards of that care which it has delegated to those farmers.

With regard to responsibility toward animals, Budiansky asserts that we have entered into an evolutionary covenant with those species of animals that are traditionally found on livestock farms, and that this imposes a duty of care on those farmers. This assertion is repeated in a variety of ways by other writers on the subject. Rollin (1993:7) for example simply states that 'traditional agriculture involved an implicit contract between human and animal... 'I take care of the animals, the animals take care of me', which was both a prudential and an ethical maxim.'

Rolston (1988:79) on the other hand discusses the matter at more depth and suggests that, in removing the domesticated species from their natural environment, or in encouraging them to leave that environment, as Budiansky has suggested, humanity has taken a duty upon itself to care for those species. 'In taking an interest in them, humans have assumed a responsibility for them.' The nature of that responsibility Rolston then suggests is that 'they ought to be treated... with no more suffering than might have been their lot in the wild, on average, adjusting for their modified capacities to care for themselves.' Scruton (1996:97) writes along similar lines when he declares that 'duties to animals arise when they are assumed by people, and they are assumed whenever an animals is deliberately made dependent on human beings for its individual survival and well-being.' He then identifies those duties *vis-à-vis* farm animals when he writes, 'the demands of morality are answered when animals are given sufficient freedom, nourishment and distraction to enable them to fulfil their lives, regardless of when they are killed, provided they are killed humanely' (Scruton 1996:99).

When one considers the issue of a livestock producer's duty to their animals in these relational terms, it is difficult to avoid this conclusion that in allowing, encouraging or enforcing animals to become dependent on human care, we take on a responsibility for them. Farm animal welfare is established as a duty, not only for the farming community but also for society at large which is dependent on them and which delegates immediate responsibility for animal care to them. Stewardship becomes an issue for the whole of society and not just for livestock producers.

Having established that this sense of duty is an integral part of our relationship with animals, especially with regard to how they are farmed, I suggest that **duty expressed as stewardship, and care expressed as stockmanship, are the two components that constitute a practical and agricultural principle of respect.** Furthermore, I suggest that this principle, understood in these terms, provides a base from which intensive livestock production systems can be evaluated ethically. This process of evaluation is the concern of the next chapter. Various issues relating to current practices and methods in the livestock industry in this country will be considered. The criteria now established will then be used as a means of determining whether they are ethically acceptable or not. At the same time, the practical and economic implications for the industry generated by the application of these principles will be discussed.

CHAPTER 6

Applying the Respect Principle to Intensive Livestock Production

6.1 Practical implications

The judgement that all extensive systems are by definition ethically acceptable and that all intensive systems are therefore unacceptable is simplistic and incorrect. As Ewbank (1994:10) has stated, 'Extensive systems are not necessarily "better" and conversely intensive systems are not necessarily "worse"... Rather we should realise that there are both good and bad units in each category, as measured by animal health, well- being and productivity. The aim should be to encourage good animal welfare, whatever the system.' Indeed, many of the welfare issues concerning farm animals are common to both types of production systems.

Moreover, pressure from consumers and welfare groups for farmers to return to more extensive production systems, as represented for example in the growing public interest in organic production methods, raises a new range of welfare concerns which have been recognised by the Farm Animal Welfare Council (1993:16). 'Whereas in the recent past welfare may have suffered due to intensification which solved the problems of profitability in a situation of limited resources, present changes may create problems of welfare a) by a change from intensification to more extensive systems e.g. hill sheep, b) by inherent disadvantages of the extensive system e.g. free-range hens.'

It has to be recognised, however, that there are specific welfare issues relating to the use of intensive systems and these are now examined in the light of the principle of respect. This principle is understood in the terms already discussed in the previous chapter, where respect is interpreted as the valuing of an animal's essence or 'animality' and is expressed in the provision of high standards of care (stockmanship) and a strong commitment to human responsibility and management (stewardship) of domesticated species.

It has also to be recognised that there are a number of ethical issues arising out of some recent developments in livestock husbandry that are not covered in this debate. For example, the rearing and production in the United Kingdom of 'exotics' such as ostriches, or the issues surrounding the cloning and genetic manipulation of farm animals, would both be outside a more general concern for commercial livestock production methods and systems. The discussion that follows omits mention of the rearing of either sheep or beef. This is because both of these sectors of British livestock agriculture still tend to use more traditional systems rather than intensive methods of production. Many beef animals are still reared in suckler herds and sheep, be they for fattening or breeding, are frequently kept in flocks that are only housed during lambing time.

Finally, I am only too aware of some of the issues that I have omitted from the discussion. For example, the slaughtering at birth of those chicks or calves which are a necessary but unwanted and unfortunate by-product of certain poultry or dairy enterprises, poses questions regarding the morality of such production systems and yet they have not been included in the discussion. This is not because this or any of the many other issues that could have been studied have been overlooked or considered to be unimportant. Rather, my purpose has been to establish criteria by which the ethical evaluation of any intensive production system can be made. I believe that the principle of respect provides the tool for this to be done and in the following examination of three specific areas of intensive livestock farming, i.e. those of egg, pig and milk production, I have tried to establish a model for the ethical assessment of any intensive livestock system.

6.1.1. Egg production

Of all the methods employed in the breeding and rearing of animals for human use and consumption, few have proved to be as emotive as the use of battery systems for laying hens. Although the commercial rearing of layers in cages goes back to the early 1900s, it wasn't until the 1930s that commercial cages were being manufactured in Britain, and it took another twenty years before the system was widely adopted amongst British egg producers. In 1995 of the 27,782 laying flocks in the UK there were only 750 battery-caged flocks, although it has to be said that those 750 flocks accounted for 91% of the national egg production (NFU, 1998:68)!

During the 1960s the welfare implications of such systems began to cause some degree of consternation. Harrison (1964:154) summarised many of these concerns by quoting words from Budichek's 'Adventures with a Naturalist'. He described how the battery chickens he had observed 'seem to lose their minds about the time they would normally be weaned by their mothers and off in the weeds chasing grasshoppers on their own account. Yes, literally, actually, the battery becomes a gallinaceous madhouse. The eyes of these chickens through the bars gleam like those of maniacs. Let your hand get within reach and it receives a dozen vicious pecks - not the love pecks or the tentative peck of idle curiosity bestowed by the normal chicken, but a peck that means business, a peck for flesh and blood, for which in their madness they are thirsting.'

Emotive words such as these cannot but influence those who have never visited a battery unit or who, having observed the operations of such a unit, interpret what they have seen in purely anthropomorphic terms. In addition, there are undoubtedly some units where the acceptance of responsibility for the well being of the hens is sadly lacking and the level of care is so poor that the birds exist in squalid and painful conditions, which are comparable to those described by Budichek. Indeed, in November 1998 Compassion in World Farming (CIWF) highlighted such concerns in a documentary broadcast on television, and from which details of the unsatisfactory conditions that their research had uncovered were passed on to the Minister of Agriculture ('AgScene' Spring 1990).

There are many other units, however, where both the environment and atmosphere of the battery shed convey a sense of contentment and well-being amongst the hens. In fact the problems related by

Budichek are not confined to battery systems. Thus feather-pecking, which can easily lead to cannibalism, is a problem known to producers using a range of production methods. Webster acknowledges, for example, that 'mortality rates of birds in large, commercial colony systems are usually at least five times greater than for hens in cages and most of these deaths can be attributed to problems of aggression, fear or stress rather than to infectious disease' (Webster, 1994:158).

Indeed, in 1996 the Ministry of Agriculture Fisheries and Food (MAFF) identified feather pecking and cannibalism as major problems in these colony systems and prioritised both as matters requiring urgent further research (MAFF, 1996:43). On this basis, rather than condemn battery systems on welfare grounds, some would actually endorse their use on exactly these grounds. As might be expected, the National Farmers Union (NFU, 1995:14) support their use and have stated that 'cages provide the best conditions for maintaining bird health by protecting against the introduction of disease and controlling its spread, for providing shelter and security against predators and for minimising aggressive behaviour and cannibalism.'

The Universities Federation for Animal Welfare (Ewbank, 1994:11) follows a similar line in suggesting that 'it could be both a welfare and economic disaster to force the present population of battery birds out into the relatively untried alternative systems.' The comment is also made that alternative systems 'are more difficult to run than battery units; they demand a higher level of stockmanship, (and) the outdoor systems seem to have a higher level of disease and the eggs cost more to produce.'

As the management of battery egg systems appears to require high standards of stockmanship and stewardship, it could be argued that under the terms of the principle of respect battery units must be considered ethically acceptable. Both qualities have already been identified as the essential components of the respect principle and as the expression of both is clearly required on battery production units and as such units appear to allow more specific care to be given to individual birds than is possible in alternative large systems, there might seem to be a case for ethical acceptance. The fallacy of this argument becomes clear, however, when one begins to apply the

196

respect principle more rigorously in terms of respecting the 'essence' of the birds themselves. Already defined in Chapter 5 as 'all that makes a thing what it is; its indispensable quality or element', one has to question whether the battery system respects this quality or whether it denies or even contradicts it.

On these grounds Webster (1994:157) finds battery systems unacceptable when he states that 'the most serious welfare problems for the caged layer are the frustration of normal behaviour and the lack of sustained fitness' as a result of the bird's restricted environment. At the present time most birds in UK cage systems are allowed approximately 450 square centimetres of space (the size of this page) with three to five birds being confined in each cage (Sainsbury 1986:160). Webster (1994:158) indicated something of the frustration and possible suffering that this causes caged birds when he stated that, without taking account of the additional space required for nesting, each bird requires 900 square centimetres to fulfil its instinctive needs. On this basis he concluded that 'the battery cage, in its current form, does not meet acceptable minimum standards for the welfare of the laying hen.' He then suggested, however, that, provided certain criteria to be met regarding the design and construction of cages, his misgivings would be put to rest.

Webster's criteria include a suitable nesting site, a perch and a greater space allowance per bird. With these provisions the bird is able to express something of its nature, 'animality' or essence and as a result there would appear to be no reason why such a battery system should be deemed ethically unacceptable. Furthermore, in allowing individuals birds greater freedom of movement such battery systems should overcome another welfare anxiety concerning the muscle and bone weakness that many caged birds suffer from as a result of the physical restrictions they are forced to endure in current systems ('AgScene' Spring 1999:22).

Webster's concern regarding the current inability of a bird caged in a battery system to fulfil its natural needs, and his readiness to condemn such systems on this basis, is reflected by many other writers, including the compilers of the Brambell Report (1965). The NFU (1995:14), however, whilst willing to acknowledge that the birds are physically deprived, makes the point that 'the extent to which the restriction of the natural behaviour of laying hens compromises their welfare has yet to be quantified and it will be extremely difficult to do so.' Despite this element of doubt, however, the general view is that because of the very limited space that each bird is allowed, current battery systems are ethically unacceptable. Indeed, on the basis that the principle of respect demands that the essence of an animal be acknowledged and valued, it would difficult to argue otherwise.

Like Webster, Fraser and Broom (1997:374) focus on the frustration a bird experiences, having been denied the expression of her natural needs. They comment that 'many of the problems for a hen in a battery cage are a consequence of things she cannot do. She cannot move freely, flap her wings, perch, build a nest before oviposition, scratch for food, dust-bathe, or peck at objects on the ground. The consequences of such deprivation are frustration, pecking at other birds and certain abnormalities of growth and body form.' In this, Fraser and Broom recognise the clear link between the frustration of the bird's natural needs and a deterioration in its welfare.

Sainsbury (1986:160) follows a similar line in condemning battery systems on these grounds. Like Webster, he acknowledges the benefits and acceptability of what he describes as 'the get-away' cage. These fulfil some of the welfare criteria set by Webster, in that the birds confined in these cages are allowed 'more space, a perch and a littered nest box'. But Sainsbury suggests with regret that a production unit using such cages would not be commercially viable.

Despite Sainsbury's conclusions regarding the economic viability of battery systems using the larger improved cage design, during late 1998 and early 1999 the Agricultural Council of the European Union seriously considered the imposition of such systems across the Union. In a letter to National Farmers Union (NFU) members on 18th June 1999, the chairman of the Poultry Committee Charles Bourns outlined details of the initial proposal that was put to the Council. This was for cages to be enlarged to '800 square centimetres per bird with a 15 centimetre per bird perch (10 centimetres above the cage floor) with a minimum cage height of 50 centimetres at any point'. Despite the NFU expressing serious misgivings about the supposed welfare benefits contained in the new proposals (NFU, April 1998) it would seem that, if they had passed into law, they would have satisfied many of the concerns of those like Webster who remain opposed to current battery production systems.

Unfortunately, however, the agreement reached by the Council in June 1999 dramatically amended the original proposals. Current systems are indeed to be banned from 2012 but, although all new cages installed after 2003 are to be in the new 'enriched' form, the requirements are nowhere near those originally proposed. Instead the new legislation requires a minimum space allowance of 550 square centimetres per bird and the fitting of claw shortening devices to all cages (EU Directive 99/74/EC); measures that fall far short of the original welfare proposals.

On the basis of the above, and applying the respect principle to egg production in battery systems, on ethical grounds one cannot but reject the system as currently practised whilst expressing grave concerns about the measures imposed under the EU's new Directive for the Welfare of Laying Hens (99/74/EC). If Sainsbury's 'get-away' cage could be made profitable, however, or if the original proposals discussed by the EU Agriculture Council in June 1999 as outlined above could become law, I would argue that battery systems could then comply with the criteria laid down by the respect principle and be deemed ethically acceptable.

One disturbing feature of the debate concerning the regulation of battery systems is that it appears to have been dominated by emotive statements on the one hand and political and economic concerns on the other. There appears to have been little reference to scientific and ethical criteria regarding the welfare of the laying birds. As has already been indicated, I believe that the financial implications of welfare legislation cannot and should not be ignored. Indeed, it is essential that the economic burden being placed on producers should receive due consideration. What these costs amount to in terms of a farmer converting from egg production in a caged system to free-range or semi-intensive systems is illustrated in Table 8.

Table 8 Cost of different systems of egg production
(Source: Fraser and Broom 1997:381)

System	Space# Square metres /bird	Relative cost (Laying cage = 100)
Laying cage	0.045	100
Laying cage	0.045 + perch	100
Laying cage	0.045 + perch + nest	102
Shallow laying cage	0.045	102
Laying cage	0.056	105
Aviary, perchery, multi-tier	0.05	105 to 108
Laying cage	0.075	115
Get-away cage (2 tiered aviary)	0.1 to 0.83	115
Deep litter	0.14 to 0.1	118
Straw yard	0.33	130
Semi-intensive	10	130 (140)*
Free-range	25	150 (170)*

Space refers in cages to cage floor area, in houses to house floor area, and in extensive systems to land area.

* Cost includes land rental

It has to be noted, however, that if society, represented by its politicians, considers that certain improvements to livestock breeding and rearing conditions are ethically warranted and are to be legally imposed, then two things are required. Firstly, politicians must have the moral courage to legislate according to their principles. Secondly, society must be prepared to pay for the increased costs that those welfare improvements impose. Such costs should not and cannot be borne solely by the producer. If they are, then this is not only unjust but is a threat to food supplies, in that the financial viability of the producer is also threatened. These financial considerations are major issues regarding all aspects of legislation *vis-à-vis* farm animal welfare and as a result will be considered more fully in Chapter 7.

To return to the EU Directive on the Welfare of Laying Hens (99/74/EC), one finds that it is not only concerned with regulating the production of eggs in battery units but also with the alternatives. These are categorised as free range or deep litter systems. There appears to be a widespread assumption that such systems are ethically more acceptable in that the birds have greater freedom to exercise their natural instincts. It would seem from this that such systems take account of the 'essence' of the bird, and that its needs are accordingly recognised and respected. When these systems are examined, however, using the criteria already established, certain ethical issues emerge. The fact, for example, that the Directive imposes new regulations regarding the allowed stocking density of birds in these systems indicates that there are already concerns about the number of birds reared per square metre in densely stocked colony systems and the welfare problems that result.

Under the new legislation, which is due to take effect in January 2002, all new deep litter or free range systems 'must have a stocking density not exceeding 9 hens per square metre usable area' (NFU Briefing June 1999). All present systems are allowed to continue with current stocking densities of 12 birds per square metre until 31st December 2012. Fraser and Broom (1997:377) point out that these allowances provide each bird with approximately the same amount of individual space as would be permissible under current battery cage legislation. It has to be appreciated, however, that the effect that this has on the bird's welfare is less than it would be in a caged environment. In these alternative systems the bird is allowed freedom to move within a much larger overall area and thus to engage in instinctive practices such as nest building, wing flapping, investigative pecking and dust bathing.

The major welfare issues with regard to free range or deep litter production systems stem from the size of the flock required for such a system to be financially viable. The statistics for laying flocks in the UK recorded by the Ministry of Agriculture Fisheries and Food (MAFF) indicate that although there were only 300 flocks of over

20,000 birds in Britain in 1992, they contained 73% of all laying birds (MAFF, 1997:13). Whilst this figure is for all laying systems, some argue that in terms of the care of individual birds, flock numbers of this size make the task of the stockperson responsible for free range or deep litter units almost impossible. As a result the welfare of the birds is bound to suffer. The reasons for this are outlined by Webster (1994:158), who describes the problems of colony systems in terms of 'intraspecies aggression, mob hysteria and the consequent fear and stress experienced by birds that are unable to maintain personal control of their physical and social environment.' His conclusion is that 'the welfare of laying hens in large, commercial colony systems is at least as bad as that of hens in the conventional battery cage.'

Flocks in which such conditions exist clearly contravene one of the basic tenets of the principle of respect in that the size of the flock makes it impossible to maintain high standards of stockmanship or to exercise fully the responsibilities required by stewardship. This in turn raises the question as to whether there is an optimum flock size for laying birds which, if exceeded, leads to a deterioration in the birds' welfare. Whilst recognising that skilled stockmanship can minimise the aggression and cannibalism of birds in a large flock, and acknowledging that the responsible use of medication can overcome any inherent disease problems, the size of the flock is still clearly crucial to the well-being of the birds. Whilst this is recognised by the Farm Animal Welfare Council in their 'Report on the Welfare of Laying Hens' (1991) they comment that there is a wide difference of opinion as to what the optimum size might be. Whilst 'some research scientists proposed a maximum group size of 40 hens...others reported satisfactory results (in regard to both production and behaviour) with up to 3000 hens' (FAWC, 1991:22-23).

The Council's recommendation was a compromise that 'maximum flock size in a colony system should not exceed 2000 hens' (FAWC, 1992:23). The implication was that in flocks which exceeded this number, adequate care and responsibility for the birds could not be exercised fully. I would also add that as a consequence the principle of respect could not be observed fully. Where colony systems have flocks in excess of FAWC's recommended figure of 2000 birds, therefore, the demands of stockmanship and stewardship require that the larger flock be split into smaller groups.

I suggest that where well-managed poultry systems exist, they are likely to be in complete accord with Rolston's criterion regarding the measure of human responsibility for domesticated animals. This states that an animal should experience 'no more suffering than might have been their lot in the wild, on average, adjusting for their modified capacities to care for themselves' (Rolston, 1988:79). In a well-managed free-range or deep litter system that does not exceed FAWC's maximum recommended size, I believe that Rolston's citerion, as well as those other criteria required by the principle of respect, are adequately met, and that such systems can therefore be considered to be ethically acceptable.

6.1.2   Pig Production

The decision of the British Government in January 1999 to ban unilaterally the use of sow stalls and tethers by UK pig producers has had a marked effect on the competitiveness of the British pig industry. Whilst UK producers have been forced to rear pigs in more welfare-friendly but consequently more costly production systems, it has remained legally permissible for retailers to import pig meat from suppliers using the very systems made illegal in this country.

This example of double standards on the part of both the retailer and the consumer who purchases these cheaper imports, sadly seems to typify much of the debate regarding farm animal welfare. As Webster has remarked, it is ironic that the livestock products that are in most demand are from those very systems that the public seems most ready to condemn (Webster, 1994:146). Indeed, it is largely because of the success that intensive methods have had in terms of the rearing of pigs and poultry that there has been the growing demand for their products, in that it is the production methods that have made those products so cheap to purchase. According to Webster (1994:146) there are three reasons for this.

'(1) Ease of handling and storage of dry, cereal-based feeds;
(2) reduction in feed costs through control of air temperature;

(3) improved control of infectious disease in densely stocked buildings by (i) vaccination, (ii) antibiotics and (iii) supply and maintenance of minimal disease stock.'

Of these three, it is the dependency of such systems on 'densely stocked buildings' that has probably caused the most public concern. In describing the production of pigs in such a unit Harrison (1964:2) wrote of 'the closely packed, inert mass of pigs on the floor of the sweat-box piggery'. She then proceeded to define the causes for concern raised by such production methods. 'The plight of intensively reared pigs in overcrowded pens containing no bedding, in dim light or in darkness, usually feeding off the floor and with no separate dunging space, is made all the more acute by the fact that pigs are naturally clean, lively and intelligent animals' (Harrison, 1964:164).

It has to be said that the extreme conditions described by Harrison would no longer be acceptable in this country. Rather, there has been a steady move on British pig farms over recent years, away from the barren densely stocked systems of the past towards 'a greater use of straw, group housing of sows (and) a progressive and continuing movement outdoors' (Chester, 1998:5). It has to be recognised, however, that as with poultry production, stocking densities and the size of rearing and fattening units can have an adverse influence on the welfare of the pigs being produced. Fortunately the situation in Britain has not developed to the extent that it has in the USA where one 'hog farm' is reportedly producing 100,000 sows and two million fatteners (Hiron, 1996:18). The environmental problems posed by a unit this size in terms of slurry disposal and water requirement are of a scale difficult to envisage. Similarly, the welfare problems raised by such a unit must be equally critical in terms of the stocking densities being employed and the stockmanship and management skills that are consequently demanded of the farm staff.

Whilst the present tendency is for large units in Britain to have 400 to 500 sows and UK environmental and planning laws make it highly unlikely that any unit will ever exceed 700 sows ('Farmers Weekly' 31st. December 1999:23) the welfare implications of continually increasing the size of British herds still need to be appreciated. The statistical evidence indicates that over recent years this pattern of growth has been taking place, with a steady increase in the number of pigs per herd in both breeding and fattening units in the UK (MAAF, 1997:12). Furthermore, the projections are for this trend to continue ('Farmers Weekly' 31st December 1999:23).

It must follow, therefore, that the demands in terms of the additional commitment and care that these expanding herds require of the stockpersons and herd managers who service them are bound to increase accordingly. As Lean (1994:166) has commented, 'The continuing process of intensification means that stockmen must increase their vigilance and awareness of the signs of problems developing in animals... It is impossible to overstate the need for sound, efficient, humane stockmanship.' The welfare problems arise when this factor is not taken into account in the managing of a unit, and the two features of the principle of respect, stockmanship and stewardship, are as a consequence either ignored or over-ruled.

Fortunately, despite the anxiety expressed above, there is much to suggest that the quality of stockmanship care and management expertise on British pig farms is of the highest standard. Indeed, this was recognised and applauded by John Godfrey, President of the British Pig Association (BPA) in an article in the journal 'Pig Industry' (April 1998:4), although he followed his words of commendation by expressing the concern that the industry required even further improvement. He proceeded to outline how this could be achieved. He made the point, for example, that UK pig producers need to ensure that they employ people who not only express a commitment to the care of pigs but who also exhibit a willingness to receive ongoing training and show the ability to benefit from it. He also argued that there needs to be an improvement in the image that the public has of both the industry and those it employs, as only then can the workforce be expected to have a pride in the work they do.

The importance of the points made by Godfrey, especially regarding the high standard of UK stockmanship, was illustrated in a report in the December 1999 edition of 'Pig World' regarding the management of a Shropshire pig unit consisting of 90 sows and 1000 finishers. The journalist commented that although the sows were bedded on rubber rather than slats or straw, they were clearly healthy and contented.

This he suggested was in many ways due to the quality of care and commitment he saw evidence of in the workforce. 'What they... have is a special affinity with their stock and with what they do, and they pay incredible attention to detail or how would they produce so many excellent pigs? I did not see one poor pig anywhere' (Walton, 'Pig World' Dec. 1999:35).

The value of both the stockmanship and management skills employed on that farm are clearly the key to the well being of the pigs being reared. The importance of these qualities has still to be recognised by society, however, and to add to Godfrey's comment, the image of the pigperson and the unit manager must be improved so that they and their skills receive the recognition that they deserve from the consumer on whom their livelihood depends. Such recognition can only lead to those concerned taking an even greater pride in their skills and to encouraging them to develop their skills even further.

Applying the criteria of the respect principle regarding the provision of the highest possible standards of stockmanship and management to the care of the pigs, I would argue that a well-run intensive unit in the United Kingdom is likely to be deemed ethically acceptable.

But certain questions remain as to whether these units exhibit any indication that the pig's essence, in terms of what the animal requires to fulfil its instinctive needs, is being valued or is even recognised. The suggestion was made by Harrison that large numbers of intelligent animals kept together in a barren and boring environment are likely to become frustrated. Such frustration is exhibited in aggressive or stereotypical behaviour. 'Pens without bedding offer little outlet for the pigs to perform their normal patterns of behaviour. Highly motivated oral behaviours often become re-directed towards pen fittings and other pigs' (Arey and Bruce, 1993:269).

Concerns such as these motivated the Health and Welfare of Pigs Bill that was presented to Parliament in July 1998 by the back-bench MP, Chris Mullin. Sponsored by Compassion in World Farming, its recommendations included 'giving housed animals appropriate bedding, an increased space allowance of up to 50% above EU and UK levels; tail docking to be initiated on a pig-by-pig basis, administered under general anaesthetic and with a vet in attendance; and a six week weaning period to be introduced' (King, 1998:20).

Whilst commending the humanitarian motives that inspired the Bill, Dr. Christianne Glossop, President of the Pig Veterinary Society argued that some of the proposals could be detrimental to pig welfare. The space requirements, for example, were described as 'a complete nonsense' in that they took no account of the sociable nature of pigs and could cause the animals some distress (King, 1998:20). Likewise, the extension of the weaning period could prove harmful to both the sow and the piglet in that 'the current modern sow is genetically and physically unable to sustain herself and a litter for an extended time. Piglets' immunity wanes as time goes on. At six weeks they are more liable to pick up diseases than they are at three to four weeks' (King, 1998:20). Glossop's conclusion is that those who drafted the bill had 'no idea of the basic psyche' and needs of pigs (King, 1998:20) and that the Bill was therefore ill-advised and inadequately formulated.

This is partially illustrated in the Bill's failure to take account of the fact that fighting and tail biting are not confined to intensively reared herds. Aggression can be a result of management problems in any production system and is a problem endemic in pigs in general. This was recognised in an article in 'Farmers Weekly' in which the comment was made that 'loose sow housing may be perceived as welfare-friendly by the public, but the reality can be somewhat different' ('Farmers Weekly', 29th October 1999:42). Despite all the efforts to reduce bullying in a unit which was moving into a loose housed system, the pigman still commented that aggression was difficult to control, and that the introduction of new sows into an established group made some aggressive behaviour inevitable.

Further arguments can be directed against Mullin's proposed Bill with regard to the recommendations for the provision of bedding for pigs. Such bedding would normally consist of straw, and whilst there has been a general move towards straw-based systems over recent years there have also been expressions of concern at the potential welfare problems that might result ('Farmers Weekly' 31st December 1999:16). Indeed, the National Farmers Union (NFU, 1995:13) have stated that 'the use of straw can require more management and labour and give rise to greater disease risks.' In discussing the bedding needs of pigs, Webster (1989:2) states that 'the ideal bed needs... to be hygienic, dry, resilient and reasonably warm'. On this basis he suggests that 'deep, clean, dry straw provides an ideal bed for weaner and grower pigs, because it is warm, dry, hygienic, resilient and a constant source of interest.' He also notes, however, that when management is poor and the straw is left fouled and wet, then it can become a welfare hazard. As a result one can argue that rubber flooring can achieve the same results as straw without the problems. In addition, when the pigs are provided with items such as tyres or rope with which they can amuse themselves one cannot but argue that such bedding systems fulfil the welfare criteria.

When the animals are in responsibly-managed units in which due attention is given to the relief of potential boredom, and in which, as Glossop suggests, the basic psyche of the animal is taken into account, I believe that the demands of the respect principle are being met.

In addition to the concerns already discussed, another feature of the current debate regarding pig production has been the gathering interest in outdoor rearing systems. Although they have tended to have a more favourable welfare image than indoor units (Fraser and Broom, 1997:366) the trend towards more outdoor production has largely been directed by the low capital investment that is required, rather than being motivated by welfare concerns (FAWC, 1996:3). There is no doubt, however, that pigs allowed the freedom to roam over an area, and given the opportunity to root and wallow, are likely to be fulfilling more of their essential animality. Such systems pay due recognition to the essence of the animal.

It would be wrong to assume, however, that outdoor rearing systems meet all the requirements of the respect principle. Indeed, as regards the exercising of care for the stock, outdoor production may well demand even more of the stockperson and manager than an indoor unit and make their task more difficult (Stark, *et al.,* 1990:11). For example, the animals in an outdoor herd are almost as susceptible to disease as those in an indoor system, whilst their condition is less easy to monitor, as the pigperson's work may be hampered by factors such as the environment, the area covered, the weather, or darkness.

There are further welfare issues relating specifically to outdoor production in that the young are more vulnerable to predators such as foxes (NFU, 1995:13) and the whole herd is exposed to the harshness of the weather. Furthermore, as with all loose housed pig production systems there is the problem of aggression. Rather than tail docking and teeth clipping being confined to the management of indoor units, a survey of 77 outdoor herds discovered that 70% of the units surveyed carried out tooth clipping and 58% docked tails (Abbott *et al.*, 1996:627) as a means of dealing with intra-herd aggression.

Table 9   Effect of housing systems on the welfare of dry sows (Source: Webster 1990:143)

	Outdoor	Indoor	
	Paddocks and yards	*Individual stalls, no bedding*	*Covered straw yards*
Thermal comfort	Very variable	Fair to poor	Good
Physical comfort	Variable	Bad	Good
Injury	Slight	Feet, 'bed sores'	Fighting
Hygiene	Fair to poor	Usually good	Fair to poor
Disease	Some parasitism – control difficult	Usually good	Parasitism - control easy
Abnormal behaviour	Slight	Severe	Slight

As a result of these concerns I would argue that, whilst there are welfare arguments in support of outdoor production units, and whilst a well-run unit fulfils all the requirements of the respect principle, an equally well-managed indoor unit can be as ethically acceptable and as welfare friendly. This is illustrated in Table 9, which compares the welfare of pigs in different production systems. The comparison

209

indicates that there are positive welfare benefits in each system, even in the now-banned stall unit, although there is a clear preference for either the straw covered yard or the outdoor system.

Further concerns regarding pig production are the employment of farrowing crates and the practice of mutilation such as teeth clipping and tail docking. This latter issue will be discussed in Section 6.2.

The debate regarding the use of stalls and tethers, is now over as far as UK producers are concerned, although it is still an ongoing welfare issue in the rest of the European Union. However, it is still common for sows about to give birth to be confined to farrowing crates. The argument for the use of these crates is that they are a welfare measure, which prevents the sow from rolling onto or overlying her newborn piglets. Fraser and Broom (1997:366) note that 'piglet survival is improved by using a warm creep area and bars on the farrowing crate which minimise the chance that the piglet will move under the sow.'

There is a welfare tension, however, inherent in the use of farrowing crates, for on the one hand they are of clear benefit to the welfare of the piglets, but conversely one could argue that they are stressful for the sow. The National Farmers Union (NFU, 1995:12) have argued for 'a balanced debate about the role of farrowing cradles in pig rearing since the welfare of both the sow and her new born piglet is at stake.' Webster (1994:150) argues, however, that a farrowing crate is 'undeniably even more uncomfortable than a dry sow stall and severely restricts the opportunity for social contact between sow and piglets. Moreover, the sow that is confined in a crate during the period prior to farrowing is unable to satisfy her powerful motivation to build a nest.' Webster's conclusion is that the use of the farrowing crate cannot be examined in isolation from an evaluation of the rest of the production system and what is being attempted in using it. He suggests for example that the accommodation for the sow and her piglets should be redesigned 'on the basis that the sow occupied a maternity suite for approximately five days, then proceeded with her piglets to nursery accommodation, the maternity suite could incorporate an existing or improved design of farrowing crate' (Webster, 1994:152).

On the basis of the criteria laid down by the respect principle, Webster's suggestions have much to commend them in that they indicate an attempt to understand the sow's needs and desires. In fact I would argue that such an attempt is basic to any expression of respect that one may have for an animal, in that it signifies that the individual concerned is endeavouring to appreciate the very essence of the animal for which they are responsible. As was indicated in Section 5.2.1 on stockmanship, this readiness to understand an animal's needs, perceptions and desires can lead to a deep empathy between stock-keeper and stock and is likely to result in a subsequent improvement in the standard of animal care. It is also clear that in emphasising the importance of post-natal care for the sow and her young Webster is again in accord with the respect principle's aim of establishing stockmanship as the essential element in farm animal care.

On this point, regarding pre and post-natal care of the sow, it is worth noting that a three-year research programme commenced in 1999 to examine the use of farrowing crates in exactly the way that Webster has advocated. Cambac JMA Research will be examining a range of farrowing systems, not with the purpose of outlawing the use of the crate, but rather to ascertain whether an alternative is required, and if so what that alternative might be ('Farmers Weekly' 25.9.99:42).

6.1.3 Milk production

In 1995 the National Farmers Union identified four main welfare issues that needed to be addressed with regard to the dairy cow (NFU, 1995 P.16). These were the welfare concerns arising out of the use of bovine somatotropin (BST), the high incidence of mastitis and lameness in UK herds and inappropriate, outdated or badly designed housing, and. As the report which highlighted these issues also praised the work of the Farm Animal Welfare Council (FAWC) and stressed that 'Farm Animal Welfare Council recommendations should guide welfare policy and practice' (NFU, 1995:5) it should come as no surprise to find that these issues largely coincided with FAWC's own specified areas of research into dairy cow welfare at that time (FAWC, 1993:11-16).

Under the European Community's moratorium on the use of BST, a synthetic form of bovine growth hormone, only two products are actually available to dairy farmers in any member state, with the usage strictly restricted to veterinary purposes. The reason for the moratorium is that in Europe the use of BST is believed to be hazardous to the health of both cattle and people. This moratorium, to be reviewed and possibly extended beyond 2000, is probably the most urgent and critical welfare issue involving the dairy industry at the present time.

Whilst the moratorium is in place across the EU, the members' farmers are protected from American dairy imports which have been produced at a lower cost as a result of their use of BST. At the same time cattle in EU dairy herds are protected from a drug which 'simply increases the capacity of the mammary gland to synthesise milk without adjusting the anatomical or physical ability of the cow to process nutrients' (Webster, 1994:172). The welfare problems for the dairy cow that result from the animal receiving injections of BST over a prolonged period are more than those of just pushing that animal's productive capabilities. Indeed, research has indicated that the adverse effects of such usage on the individual animal are equally apparent in the increased occurrence of mastitis and lameness, impaired conception and tender injection sites (FAWC, 1997:95).

It is possible that some observers on the other side of the Atlantic may consider these fears to be groundless and even believe that they have been deliberately exaggerated within Europe to exclude American dairy products from European markets. In a very candid appraisal of one of their own BST products, however, even Monsanto have warned that there could be some unfortunate side effects for treated animals. These include 'reduced pregnancy rates, increases in cystic ovaries and disorders of the uterus; small decreases in gestation length and birth weight of calves; increased twinning rates; higher incidence of retained placenta; increased risk of clinical and sub-clinical mastitis; increased frequency of use of medication in cows for mastitis and other health problems; increased body temperature; increased numbers of cows experiencing periods 'off-feed'; increased numbers of enlarged hocks and lesions (e.g. lacerations, enlargements, calluses) of the knee; disorders of the foot region; reductions in haemoglobin and the hematocrit values; injection swellings, which may remain permanent' (cited by Mepham, 1999:389).

When a drug manufacturer warns prospective users that such hazards are inherent in the application of one of their own products, one is frankly surprised that any herdsperson would consider risking the health and welfare of their animals in this way. Indeed, I would argue that anyone who knows of these hazards and who still proceeds to use this substance to increase an animal's milk production actually offends the principles of care and duty to their stock. Their behaviour is directly opposed to all that is implicit in our understanding of respect for an animal, as it would seem that the animals concerned are being deliberately placed in a system designed to harm them.

The ethical arguments raised by this issue go even deeper. I would argue that a livestock production policy which encourages farmers to push an animal's production capabilities beyond the point at which it is able to maintain that level of production, must of itself be morally questionable. As Webster (1994:172) noted, 'injections of BST are likely to intensify the cow's conflict between the problems of hunger, digestive overload and physical exhaustion.' A technology which imposes such welfare burdens on an animal shows little evidence of respect for that animal in terms of valuing it for its essence, for what it is. Rather, that animal is seen as an organic machine, which can be pushed to and beyond exhaustion point with impunity. The concepts of empathy and care as laid down in the principles of stockmanship would appear to be completely absent.

Furthermore, whilst stockmanship expertise may lessen the impact of the imposed conditions on the cow, in a system which takes no account of the animal's needs and feelings, it will not alleviate them. In fact the opposite may well be the case, in that a stockperson who believes it is important to have a strong empathy with their stock is likely to be frustrated, offended and angered by a system that treats animals in this way. As a result, their own standards of work may be adversely affected, and the well-being of the cows may further

213

adversely affected, and the well-being of the cows may further deteriorate. Alternatively only those who lack stockmanship ability will be employed on such units and the overall effect will be the same with a decline in the welfare of the herd.

A further point which dairy farmers may well take into account in this debate is that there is some evidence to suggest that it is not just financial folly to push cows beyond their natural production capabilities, but even to push them to those capabilities. This was the issue raised by Ferris (1998:5) when he posed the question 'if a dairy cow of high genetic merit has the potential to produce 9000 litres does that mean it is essential that she does produce at this level? More often than not in practical farming situations the optimum profitability of an enterprise is achieved at optimum and not maximum outputs.' From this it would seem that the application of a drug which effectively pushes an animal beyond its natural maximum milk yield must raise questions of a financial as well as a welfare nature. In the long-term, therefore, it may well be that the injection of BST into dairy cows will not only be considered ethically unacceptable but financially unacceptable as well.

Largely as a result of the welfare concerns which surround the use of BST, there is ongoing political pressure within Europe for the moratorium to be extended, and every indication that it will be. Indeed, it would seem ironic if the use of a drug designed to increase milk yields was permitted in European dairy herds, when there is currently a quota regime in place across member states to curb any over-production of dairy produce (Johnson, 1996:58).

Monsanto's warning regarding BST usage indicated that the drug could effectively increase the incidence of certain welfare problems that are already endemic in the UK dairy herd. Two that were mentioned in the drug manufacturer's list were mastitis and lameness, both of which were problems identified by the NFU and FAWC as matters for real concern in today's dairy industry. As a result it would be easy to imagine that they are phenomena unique to modern dairy systems, especially as these systems can exasperate both complaints. The truth is, however, that both are as old as the domestication of the dairy cow, both can be found in all types of dairy production and both can be alleviated if not eradicated by a combination of good housing and good stockmanship.

I began my farming career in the 1960s on a well-run dairy farm, milking between sixty and seventy cows through a byre with a bucket milking system and an overhead vacuum line. It was far removed from factory farming, with the cows allowed out all summer and over-wintered in a straw yard and not cubicles. Much trouble was taken over cleanliness and the exercising of high standards of stockmanship, but the major welfare problem in the herd was mastitis. Indeed, the textbook that was my main source of inspiration at the time, 'The Principles of Dairy Farming' (Russell, 1969), emphasised the disastrous impact that mastitis can have on individual animals and the wider herd. Russell describes staphylococcal infections as 'being very destructive of udder tissue' to the extent that on some farms it could become 'a serious source of loss' (Russell, 1969:163).

This is not an attempt to discount the significance of mastitis in current dairy systems, either in terms of the distress caused to the animal, or the financial penalty experienced by the farmer. It should be noted, however, that there have been dramatic reductions in the incidence of mastitis in UK dairy cows over recent decades (FAWC, 1997:33). 'Somatic cell counts (SCC) provide a broad indication of the general level of udder health within the herd and have fallen from almost 600,000 cells/ml in 1971 to about 170,000 in 1997 (FAWC, 1997:33). This has been attributed to a number of factors from improved standards of stockmanship to milk buyers imposing financial penalties on producers whose milk has a high SCC and paying bonuses for milk with a low SCC (FAWC, 1997:33).

There is still ground for concern at the levels of infection that are present in UK milking herds. Mepham (1999:390) gave voice to such concern when he noted that the UK herd suffered 'approximately 40 cases of clinical mastitis per 100 cows per annum, each case costing the farmer between £40 and £186.' Furthermore there can be no doubt that mastitis is a painful condition for the animal. 'Caused by the action of disease organisms which have entered the udder via the teat orifice' and leading in severe cases to the swelling of the infected udder, a raised temperature and even in some cases to death (Leaver,

1994:42) it remains a major welfare issue. My intention is to emphasise the point that this particular infection has long been a major welfare issue in the dairy industry and is not a phenomenon unique to modern intensive dairy systems and related methods of husbandry.

As Sainsbury (1986:19) has noted, the 'chances of mastitis increase considerably if the cows are dirty, if their accommodation promotes the viability of the organisms that can cause disease, and/or if the end of the teat can be injured.' As a consequence it is easy to identify production systems and milking practices with an increased likelihood of mastitis infection. Dirty straw yards that are infrequently restrawed can be more infective than a well-managed and regularly cleaned slurry based system. A milker who washes an udder with a dirty cloth out of a bucket of cold water, as often happened in the past in byre milking, spreads more infection than their modern counterpart who uses a spray of water and dries each udder with a new paper towel.

There are some issues, however, concerning mastitis infection, that are directly linked to current practices in milk production. Sainsbury (1986:20) notes, for example, that one problem with cubicle housing is that the beds 'quite often become dirty at the rear area and there is a tendency for a lot of slurry to accumulate in the passageways, so that the cows may become very dirty. That ubiquitous organism, the bacterium *E. coli*, is given favourable conditions to multiply and so *E. coli* mastitis, often called environmental mastitis, has become a major disease in dairy cattle.' In this case the production system is clearly identified as being at fault and causing distress to the cow.

I would argue, however, that it is not necessarily the system that is wrong, but the quality of the management and stockmanship that are being exercised in that system. In Sainsbury's scenario, for example, there is little evidence of respect for the essence and therefore the needs of the cow, nor consequently of a management style that pays attention to those needs in matters such as cleanliness or recognition of the importance of comfortable bedding. If that respect were present, and if it were expressed in a greater sense of responsibility towards the animal, which in turn resulted in a commitment to the highest standards of stockmanship, I suggest that the welfare problems would be overcome. As a consequence, and with these safeguards in place, I would then argue that the system of having cows in cubicles could be considered to be ethically acceptable.

In practical terms, such responsible management would ensure that slurry was regularly cleaned away, and that cubicles were well designed, comfortable and large enough for the cow to lie down in without the animal having to stretch out into the walk-way or slurry channel. Such management would require that the cubicles were well maintained and that they were kept clean and dry. Due regard to the breeding of the cows would ensure that the animals would have well-shaped udders and not the pendulous appendages that so often lead to teat damage. Committed stockmanship would ensure that the milking process was carried out carefully, thus minimising any risk of teat damage, which would lead to teat infection, whilst regular attention to such matters as the drying off procedure for cows preparing to calve would further minimise any such risks. I argue as a result that adherence to the criteria laid down under the respect principle actually ensures not only that a cubicle system is a sound and efficient method of milk production, but that when managed properly is an ethically acceptable system for farming livestock. I maintain, therefore, that the welfare problems found on such units can be minimised and overcome by good stockmanship and responsible management.

In the same way I maintain that the other causes of high levels of mastitis infection in modern dairy production systems can be alleviated or removed if the manager or herdsperson pays due regard to the criteria of the principle of respect. In defining the reasons for what he believes to be the unacceptable levels of mastitis in UK dairy herds, Johnson (1996:57) suggested that 'the pressure to produce more and cheaper products is ever-present, and modern dairy cattle are increasingly susceptible to mastitis as selective breeding and dietary supplements are used to boost the amount of milk each cow yields.'

Part way through 1999 UK milk producers began to experience a dramatic decrease in the price that dairies were willing to pay for raw milk. As the downward spiral in prices has continued, so Johnson's words of warning have become increasingly relevant. Producers have been put under the very financial pressures that encourage the production methods that he identifies as major causes of high levels of mastitis infection. When the producer is committed to the principle of respect, however, all effort will be maintained to ensure that the animals' nutritional needs are fully met, and that a well-managed and responsible breeding programme is sustained. To do anything less would be a denial of one's respect for the essence of the animals in the herd. At the same time any abdication of responsibility for that herd in terms of a decline in standards of stockmanship, even in the face of financial hardship, would be considered unacceptable. Indeed, personal pastoral experience, as an agricultural chaplain, of supporting a number of dairy farming families during the current crisis has led me to believe that respect and care for the stock are often given such a priority that some producers put the welfare of their animals before the needs of their own families!

An indication of this level of commitment and care was provided in the NFU report 'The Human Cost of the Farming Crisis: Audit for Action' (September 1999). The National Pig Association (NPA December 1999:12), in their summary of the report, described how farmers were continuing to meet the veterinary and nutritional needs of their stock by cutting back on family expense. 'One-half of respondents had not taken a family holiday in the previous two years. Over one quarter had cut back on entertainment and home improvements while twenty per cent had reduced spending on new clothes and seventeen per cent on white goods (electrical goods). But as many as eleven per cent of families had also been forced to cut back expenditure on food.'

Just as concerns over levels of mastitis infection can largely be countered by a dairy farmer adhering to the principle of respect, so I would suggest that some of the welfare concerns regarding the incidence of lameness in UK dairy herds can be dealt with in a similar manner. Foot problems have long been a welfare issue on British dairy farms to the extent that in the 1960s Russell (1969:39) commented on its 'severe consequences on milk yields'. The Farm Animal Welfare Council (FAWC, 1997:24) suggests, however, that as with mastitis lameness is now 'at an unacceptably high level.'

A number of reasons for the current prevalence of foot problems in dairy cattle have been identified. Thus Webster (1994:143, 172) suggested that the increased use of silage has exacerbated the problem, as the feet of the cows have become infected as a result of the animals standing in the 'run off' effluent from wet silage. In addition, as wet silage causes wet faeces the cow will possibly be more prone to foot infection if it is forced to stand in wet acidic slurry for any period of time. Where this is the case, one might respond by suggesting that the situation could be resolved if more care was paid to making drier, less acidic silage, to managing the feeding system and possibly to the housing of the animals. Due regard to the responsible management of stock such as is required by the criteria established under the principle of respect would seem to solve the problem.

One could only wish that the situation were as uncomplicated as this, and as easily resolved. Webster (1994:172) indicates the true severity of the problem when he states that 'inspection of cull cows at slaughter reveals evidence of past or present foot damage in nearly all animals. In other words the UK dairy industry is living with a painful, crippling disease with a morbidity rate close to 100%!' He also addresses the complexity of the condition, when he examines 'some of the main factors involved in the complex aetiology of foot lameness' (Webster, 1994:172) as outlined in Table 10.

Table 10   **Factors predisposing to lameness in dairy cows** (Source: Webster, 1994:173)

Factor	Examples
Breeding	Hind leg conformation, foot colour
Nutrition	Direct: laminitis on starchy feeds Indirect: wet, acid grass silage, slurry
Housing	Poor cubicle design. Hard, wet, slippery concrete
Management	Poor foot care.
Behaviour	Prolonged standing of submissive cows
'Stress'	Physiological changes about the time of parturition

If the respect principle is applied to each of the factors in the above table the assertion can be made that each example is the result of that principle being ignored. For instance, there appears to be a repeated failure to take account of the essence of the highly productive and very specialised animal that is the modern dairy cow. Indeed, there seems to be a lack of understanding and appreciation of the animal's most basic needs and nature on the part of those entrusted with that animal's care.

Most UK herds consist of Holstein-Friesians (Carter and Stansfield, 1994:41), a breed which requires the herd manager to pay particular attention to matters such as the genetic quality of the stock, the nutritional needs of the cows and the reduction of any physical and environmental stress that the animals may suffer. Indeed, these are some of the very factors that are cited by Webster in the above table. Where this attention is lacking, then it should come as no surprise that problems such as lameness occur.

The examples listed in Table 10 also suggest that where there is a high incidence of lameness there is also likely to be an absence of the two essential elements of the principle of respect, i.e. responsible management and committed stockmanship. This is indicated in each example in the table, from poor attention to breeding details through to the retention of poor stock housing and the lack of adequate foot care. From this I argue that the converse is also likely to be true and that where these two elements are given due regard then foot problems in dairy cattle can and will be alleviated. Webster (1994:175) clearly arrives at a similar conclusion when he suggests that 'the problem can be kept under control through the application of some classic principles of good husbandry and some new knowledge relating, for example, to foot trimming and management at calving time.'

One area where new knowledge and practice can have an impact on the incidence of foot problems in dairy cattle was reported at the first national conference on the subject which was held at the National Agricultural Centre, Stoneleigh in March 1999 ('Farmers Weekly' 26th March 1999:50). Various 'locomotion scoring systems' were presented as means of identifying the prevalence of lameness within a herd. The comment was then made that the use of these systems 'provides the information which will form the cornerstone of any prevention programme and singles out individuals requiring treatment and relief from suffering.' I suggest that the skills required for the managing of such a system are intrinsic to all that has already been defined as stockmanship. The system requires that the stockperson has the knowledge and ability to understand the nature and therefore the behaviour of the animal and that they combine it with an empathetic concern for the suffering of that animal. This in turn enables that person to observe and analyse the results of that observation, and results in the required help being given to the animal concerned. Criteria arising out of the principle of respect are applied and the animal's suffering is thus relieved.

At the same conference one speaker offered a 'seven-point plan for lameness control' ('Farmers Weekly' 26th. March 1999:51). Again an analysis of the measures being proposed indicates that the emphases of the principle of respect in terms of responsible management and committed stockmanship are highly relevant to the working of the plan which requires:

- 'An annual assessment of lameness by the vet or contractor who would also carry out locomotion scoring.
- Access to a good foot crush in a well lit area.
- Prompt treatment of clinical cases.
- Routine foot examinations at drying off.
- Footbaths for controlling infection or hardening feet.
- Assessing cubicle comfort.
- Making dietary and environment changes gradually.'

Besides requiring high quality management and stockmanship, however, this seven-point plan also raises the issue regarding the suitability of the housing system. Cubicle comfort is an important factor in dealing with lameness, and the size of cubicles, the type and condition of the flooring, especially when it is made of concrete, have already been mentioned as causes of foot problems in dairy cows. The Farm Animal Welfare Council (FAWC, 1997:6) recognised the role that inadequate or inappropriate housing can have in exasperating these problems. 'Lameness can result from a variety of factors e.g. cubicles which are too short; rails or kerbs inappropriately positioned or designed; inappropriate floor surfaces in housing areas which cause damage or slipping or poorly designed housing.'

The Council's concerns regarding the housing of dairy cattle are not restricted to the issue of lameness, however (FAWC, 1997:6). Recent genetic developments, combined with the introduction of more Holstein influence into the UK dairy herd, have resulted in the tendency for the modern dairy cow to be larger, and especially longer and taller, than its predecessors. The consequences of such cows being housed in cubicle systems that were designed and constructed ten or twenty years ago are that the animals suffer discomfort and even injury when they try to lie down or to stand. This in turn causes them to spend more time standing out of the cubicle, which again causes increased distress and discomfort. FAWC's recommendation (1997:20) is that 'modifications should be made to existing cowsheds if ventilation is poor, if cows are too large for the stalls or there are other design faults.' However, there is a deeper issue here.

The design of appropriate housing for any livestock must take account of a variety of factors, which whilst they include efficiency of operation and economy of scale must also appreciate the importance of other influences such as the 'nuisance factor'. For example, local communities are often anxious about the visual impact that livestock housing may have on the landscape. They may express concern at the possible nuisance 'that may be created by large conglomerations of livestock, arising chiefly from the smells of the stock and their bedding, but above all from the excreta, especially if this is produced in slurry form' (Sainsbury, 1986:7). The designer has to appreciate and work with the concerns and constraints imposed on him by all who are likely to be influenced by the housing design.

I would argue, however, that the demands of human society are not the most important part of this process, but rather the attention given to the needs of the animals themselves. Sainsbury (1986:10) appears to agree when he suggests that any animal housing system should 'give the animals generous space allowances, keep them in smallish groups and... not rely on artificial environmental control.' His attention is focused on the needs of the animal and he is motivated by an understanding and respect for the nature, the essence, of the animal concerned. This is reflected in his concerns regarding battery systems for poultry (Sainsbury, 1986:169), as he attempts to understand the behavioural needs of the birds and on the basis of that understanding tries to assess the most appropriate form of housing for them.

Sainsbury's concerns exhibit an adherence to the principle of respect and a realisation that the principle provides the basis for designing and building acceptable and thus appropriate livestock housing. Webster (1997:63) reflects a similar understanding when he states that an animal's physical comfort depends on 'a suitable site to rest and sleep in any position it fancies, and sufficient space to groom itself and indulge in modest, relaxing exercise, like limb stretching and wing flapping.' Such modest requirements are fully in accord with Rolston's demands (1988:79) that no domesticated animal should be subjected to conditions worse than those that they would experience in the wild. Furthermore, they indicate an appreciation and valuing of the animal itself, a respect for its basic needs and nature. This is illustrated in Webster's comment that 'we who are responsible for animal husbandry could often make life more comfortable for our animals and save ourselves money by first asking animals what they want and then providing them with a habitat that allows them to express their preferences in a constructive way' (Webster, 1997:73).

The practical implications of such an approach are illustrated in Table 11 (Webster, 1997:62) in which Webster identifies the major environmental requirements for animals. Moreover, I suggest that his obvious understanding of their perceptions and needs also serves to illustrate the extent to which he personally values and understands their essence. In his concern for all that an animal requires of a housing system, Webster indicates again his ability to appreciate and thus respect the very essence of an animal. At the same time he draws attention to one of the major welfare issues identified by both the NFU (1995) and FAWC (1997) concerning the modern dairy cow, in that both organisations mention the problems that can arise from the use of inadequate or inappropriate housing. Indeed, reference has already been made to the suffering and discomfort that can be caused to a dairy cow which is accommodated in outdated cubicle designs that are now too small to accommodate the Holstein-Fresian animals that make up the larger part of the UK's dairy herd (NFU 1995:16). Due regard to the principle of respect will ensure that such problems do not occur, as the use of such housing systems will be seen as an offence to the criteria of responsible management and committed stockmanship.

Table 11   **Major environmental requirements of animals** (Source: Webster, 1997:62)

Comfort:	Thermal – neither too hot nor too cold.
	Physical – a suitable resting place; space for grooming, limb stretching, exercise.
Security:	Of food and water supply.
	From death or injury due to predation, aggression, floods, etc.
	From fear of predation, aggression, etc.
Hygiene:	To avoid the discomfort of squalor and the danger of disease.
Education:	To acquire the knowledge necessary to achieve comfort and security during independent adult life.

The accommodation requirements for a dairy herd have been identified by FAWC (1997:13) as being 'of a type that provides each cow with unhibited access to wholesome feed and water; a comfortable, bedded, well-drained lying area; shelter from adverse weather; and space to move around and interact socially (which includes space to allow a subordinate animal to move from a dominant one).' One could argue that most dairy units were originally built with such criteria very much in mind. The problem is that over the years the size of both the herds and the individual animals has increased (NFU 1995:16) and some dairy units now provide inadequate housing for the herd.

In the present financial climate, the practicalities are that whilst producers may want to update and upgrade their stock accommodation as the situation demands, financial constraints prevent them from doing so. In such cases there is an undoubted tension between what a farmer believes and desires regarding required standards of stock management and care and what that farmer is actually able to practice. For milk producers such issues are particularly pronounced. For example, pressures to increase herd profitability ensure that genetic improvements in terms of a cow's milk productivity will continue to result in animals with different physical and environmental needs to those of their predecessors. Those pressures will also result in a continued increase in herd size as more cows obviously mean more milk and therefore improved income. At the same time, however, the difficult financial climate prevents the farmer from making the necessary adaptations to stock housing to allow for those physiological changes in his cows and the increased size of his herd.

Whilst it may be easy, therefore, to condemn the condition of some outdated dairy housing, due recognition of the problems facing the dairy industry demands that a greater degree of sensitivity be shown. Until such time as milk production again becomes profitable, the inherent qualities of stockmanship and herd management exhibited by the majority of dairy farmers must be encouraged. One would hope that when the financial climate improves any necessary adaptations of housing system will then be affordable and carried out.

A similar response is needed in the applying of planning regulations to buildings designed for livestock production systems. As has been frequently recognised, many of the most attractive landscapes in Britain are not natural but are the result of generations of human activity in the countryside, moulding and changing the landscape through farming practices (Newby, 1988:5). Indeed it has been argued that the quintessential feature of the English landscape is the patchwork countryside created by its network of hedges and walls, many of which will have been erected during the agricultural enclosures for the containment of livestock (Rackham, 1995:190).

In the same way, the traditional stone built field barns of the Yorkshire Dales, which are equally admired for their unique contribution to the landscape, solely exist because of past livestock rearing practices (Hartley and Ingilby, 1981:62). Or again, the fells that provide the backdrop to these barns and whose beauty attracts thousands of visitors each year are the result of hundreds of years of well-organised grazing management (Hartley and Ingilby, 1980:11).

However, farming systems have dramatically changed over recent decades, along with our understanding of an animal's welfare needs. Small, hedged fields are no longer appropriate to the demands of a modern dairy farm, whilst ancient stone barns on the whole are no longer adequate for the housing and therefore welfare needs of today's stock. In addition, if a livestock farmer is to remain financially viable, especially in the current agricultural climate, he must be allowed to utilise efficient modern housing and production systems. Furthermore, I believe that the exercise of respect for the animals, in terms of responsible management and committed care, will necessitate the redundancy of some of these landscape features from the past. The maintenance of the landscape itself, however, will still require the grazing regime provided by those farms (Bonham-Carter, 1971:178, 199). The continued existence of the scenery which so many value depends on responsible livestock management and yet, ironically, the contemporary housing needs of animals is considered by many to be an intrusion into the landscape that should be avoided at all costs.

There is a tension in this instance between human aesthetic appreciation and physical recreation on the one hand and demands for high animal welfare on the other. I contend that in matters such as this, respect for animals must be predominant. But there is a further matter to be taken into account - the welfare of the local community. If the social sustainability of an area depends on a viable livestock economy, and the endurance of the natural environment requires the continuation of managed grazing, it is clear that livestock production must be encouraged and supported. This can only happen when the application of planning regulations regarding farm buildings is carried out with understanding, sensitivity and flexibility. Measures can be taken to hide or disguise such buildings when they might be considered inappropriate in a particular setting, or to ensure that they blend in with the landscape. In supporting the welfare of the local community the welfare of the animals farmed by that community should also be accorded due regard. This regard should be expressed in the provision of high quality stock housing and the care that the principle of respect demands.

6.2 Animal mutilation

As I indicated above, controversy often results from a tension between the interests of human aesthetics and the demands of animal welfare. Similar conflicts surface when the practice of carrying out certain surgical procedures on farm animals in the interest of husbandry systems is examined. In some instances, for example, there appears to be a conflict between human convenience on the one hand and respect for the animal on the other. Where the judgement has been made that this is the case, and that an animal is suffering physical mutilation purely for the convenience of the stockperson or the efficiency of the production unit, then British law has tended to outlaw that practice.

The docking of cows' tails, for example, still carried out in a number of countries such as New Zealand and Ireland, has been prohibited in this country since 1974 when it was made illegal under Section 2 of the Agriculture (Miscellaneous Provisions) Act 1968 (HMSO, 1968). In this operation a docked cow suffers a painful surgical procedure and loses some of its ability to cope with flies during hot weather, largely to save the milker the inconvenience of having to cope with a hard, wet and dirty tail during milking. As a result British legislators judged that the welfare of the animal outweighs the comfort of the stockperson and made the practice illegal.

To imagine, however, that human convenience is the only reason why a farm animal is subject to physical mutilation is to misunderstand the intention of many stockmen when they carry out these procedures. Contrary to certain perceptions, many operations are actually performed out of a genuine desire to spare an animal further suffering at a later stage in its life. In fact, rather than being a practice unique to intensive husbandry systems, mutilations in one form or another have long been carried out on traditional livestock farms for this very reason. This fact seems to be lost on those who would condemn and outlaw all husbandry practices that involve mutilation.

This judgement would seem to be confirmed in the report on this subject from Compassion in World Farming (Stevenson, 1994:4). Reference is made to 'the widespread practice of mutilating farm animals. Lambs are castrated and tail-docked, hens and turkeys de-beaked, and the lower half of piglets' tails is often amputated. In the case of some intensive systems, animals develop abnormal behaviour in response to the limitations of the system. Factory farming's answer is not to modify the system to make it more responsive to the animals' needs. Instead they are mutilated to make them 'fit' an appropriate system.' Whilst I agree with the implied criticism of the practice of mutilation when it is a used as a means of adapting the animal to a production system, I suggest there are occasions when the operation is carried out to protect the animal from suffering. Indeed, this is often the case in extensive as well as intensive systems.

Working on a dairy farm, there were many occasions for example when I would debud a calf's horns. The operation involved the young animal first being injected with a local anaesthetic to deaden the pain, a measure required by law, and then the immature horn bud was destroyed. This was achieved by applying a specially designed hot iron to the bud, although other methods such as the painting of the bud with dehorning collodion could have been equally effective. As the anaesthetic wore off, so the calf would exhibit some sign of distress, although this soon passed. A short and relatively painless operation was carried out to prevent more serious damage being done to other cows at a later date if those horns had been allowed to grow.

The procedure was motivated out of a concern for the needs of the animals, so that at a later stage in their lives they would not be able to harm each other. It was based on a sense of duty for the members of the herd, and was carried out carefully and competently with due regard for all the demands of stockmanship. It was never a pleasant job and one can only hope that the gene that causes natural polling in certain breeds of cattle will eventually be identified so that the need to debud calves will be consigned to agricultural history.

On the grounds outlined above I believe that it would be wrong to condemn debudding as being disrespectful towards the animal, indeed I would suggest that the opposite is the case. Most cattle rearing systems used in this country require that the animals are housed for at least a part of the year. To do otherwise would be to ignore their need for warmth and shelter from inclement weather and would consequently contravene the responsibility aspect of the respect principle. To manage housed cattle one must take into account the possibility that, inadvertently or deliberately, horned animals are likely to harm each other (Russell, 1969:95). As a result I would argue that the debudding of a calf's immature horns bears evidence to the exercising of responsible management and caring stockmanship on the part of the person carrying out the operation.

In the same way I would argue that whilst the ear tagging of sheep, cattle and pigs involves some degree of pain, and might be considered by some to be a mutilating operation, it can still be justified according to the criteria of the respect principle. Indeed I believe that the practice encourages high quality stock management and care and the production of safe, healthy meat. Ear tagging ensures that full records of stock sales and movements can be kept and the complete traceability of livestock can be assured. In this way the procedure aids and encourages overall responsibility for the animals and is a major asset in the exercising of stockmanship skills and care.

Similar arguments have been made for other husbandry practices that involve mutilation. Castration, for example, has long been used in livestock husbandry systems, not only as a means of preventing uncontrolled breeding but also as a way of arresting aggression and sexual frustration in young male animals kept in a herd or flock. Strict regulations and detailed guidelines now ensure that procedures are carried out carefully and sympathetically (Webster, 1994:193). Whilst there may be a debate as to the best procedure to use, and subsequent discussion concerning the merits of a surgical blade or a rubber ring, an understanding and respect for the essence of the male animal, especially when that animal is not to be used for breeding, would suggest that castration is a sensible livestock management tool.

As with horn debudding there is evidence to suggest that the operation can be painful (Stevenson, 1994:15). It can be argued, however, that the pain experienced for a short time at this early stage is greatly outweighed by the diminution of welfare problems that the animal might have suffered at a later point in its life if the operation had not been carried out. This cost/benefit equation clearly supports the use of this procedure as a management tool and a welfare measure.

Having witnessed the aggressive behaviour of a group of young rams, for example, which resulted in one being killed by the others, I have no hesitation in asserting that in certain instances the dictates of caring stockmanship demand that castration be carried out. Indeed, this approach is recommended by FAWC (1981). Whilst 'the Council considers that, on ethical grounds, the mutilation of livestock is undesirable in principle... It acknowledges that some mutilations must be accepted when scientific knowledge and practical experience have shown that they are necessary for husbandry reasons or that greater suffering may be caused if they are not carried out.'

It is worth noting that the over recent years the public's taste in sheep meat has moved away from mutton in favour of lamb. As a result young male sheep are sold for slaughter before they reach the age of sexual maturity, which means that the possible problems of sexual frustration and aggression in young rams are unlikely to arise. Castration as a husbandry tool therefore becomes superfluous. Similar trends are evident with regard to beef, leading again to a likely a decline in the use of castration in the national beef herd.

As stockmanship has been identified as an integral feature of the respect principle, it is worth noting that, in advising the government on matters relating to the castrating of male animals, FAWC (1981) expressed concern at 'the level of competence amongst stock-keepers.' The Council's subsequent advice to the government was very much in line with the recommendations made earlier in this thesis, in the discussion concerning the principles of stockmanship *vis-à-vis* the provision of adequate training for all those who are responsible for the care of stock. They advised 'that further emphasis on training and certification in tasks involving the mutilation of farm animals is essential' and recommended 'that the organisations concerned should be asked to explore how training and certification in the relevant techniques could be given much greater prominence' (FAWC, 1981). Following the principles already established above, one can only reiterate the importance of training for all stock-keepers in the skills and principles of stockmanship.

Of all husbandry practices involving mutilation, the tail docking of sheep is probably the most widespread (Stevenson, 1994:14) and

230

arguably the most necessary. The suffering caused by fly-strike is such that afflicted animals experience great distress and in extreme infestations often die. The flies are attracted to wet or dirty fleece. Long tails often provide an ideal environment for the laying of eggs and the hatching and developing of the resulting maggots.

The most effective means of protecting the flock is good shepherding, which requires the exercising of effective flock management and committed stockmanship. Even when these qualities are in evidence however, fly strike can still occur. Webster (1994:192) noted that 'tail docking undoubtedly reduces the risk of sheep contracting maggots through blowfly strike and is therefore a positive contributor to welfare.' Similarly, whilst he emphasised the importance of competent stockmanship and recognises the damage that can be done to an animal by excessive docking, Henderson (1990:348) also appreciated the welfare value for sheep of tail docking.

When it comes to the same procedure being carried out on piglets, however, the issues are not so clear-cut, especially when the principle of respect is applied. Although the political attempt in 1997 to have the practice banned was unsuccessful, strict regulations remain in place to ensure that the procedure is only carried out when there is a likelihood of an outbreak of tail biting in a herd. In supporting the sufficiency of the current legislation, Dr. Christianne Glossop of the Pig Veterinary Society commented that 'No one would condone the routine tail-docking of any species. However in the light of the harmful consequences of tail-biting, which can include severe injuries and even spinal abscesses, the Society believes that tail docking is an acceptable procedure' ('Pig Industry' April 1998). Stevenson (1994:5), speaking on behalf of Compassion in World Farming (CIWF), expressed equal concern at the suffering caused by tail biting and cites Arey, who described how wounds can subsequently 'become infected, resulting in abscessation of the hindquarters and the posterior segment of the spinal column. Secondary infection may occur in the lungs, kidney, joints and other parts.'

Stevenson argued, however, that whilst tail biting can be the cause of great suffering in a herd, the solution is not to dock the tail but to resolve the reason for the behaviour. This, he suggested, can be

231

caused by a number of factors including 'diet, poor atmospheric environment (i.e. a build-up of gases such as ammonia and carbon dioxide) and poor housing' (Stevenson, 1994:5). Similarly, Sainsbury (1986:82) provided an even more extensive list of causal factors relating to the habit of tail biting: 'overcrowding, housing pigs of vastly different size together, poor ventilation, draughts, over-heating of the pigs, an absence of bedding, insufficient food, nutritional imbalances, low fibre in the food, diarrhoea, parasitism, skin infection, abrasions or any trauma on the pig leading to bleeding, excessive light or dark, and finally certain inherent characteristics. This is not an exhaustive list but what it does show is that tail-biting may be said to be most probably due to some effect in husbandry.'

The respect principle requires that stock should be managed responsibly and a defective production system that causes animals to be violent towards each other as a result of those defects cannot be considered ethically acceptable. In addition, the partial resolution of the problem by mutilating the animals rather than by dealing with the cause of the problem must be equally unacceptable. The demands of respect for the animal in terms of sound management practice and caring stockmanship are clearly not being fully met.

A further concern regarding the practice of tail-docking piglets, is that work carried out by CAMBAC JMA Research, an independent pig research group, has shown that although tail docking lessens the incidence of tail-biting it does not remove it. Two groups of pigs were compared, one group retained their tails, the other had them docked. In the second sample incidences of tail biting were 9%, whilst in the first they were 3% (King, 1998:20). The operation clearly made a difference, but equally it did not fully resolve the problem. From the evidence of the research King concluded that 'you don't eliminate the problem by docking, but it does make a difference. More research is needed to find the best system to eliminate tail biting.'

Until research establishes the reasons for this behaviour, the principle of caring for the stock requires that tail-docking should continue despite the fact that the practice is questionable because it does not deal with the factors causing the problem. CAMBAC's work raises a further concern in that the incidence of tail biting in both herds was so small (i.e. 9% and 3%) that one wonders if it warrants the tail docking of the whole herd. Webster (1994:112) expressed similar concerns. He argued that although a prohibition of tail docking would be detrimental to pig welfare 'tail-biting is a symptom of an unsatisfactory husbandry system which implies that tail-docking is a crude and ineffective solution to a problem that can be addressed by other more humane means.'

He made the same judgement regarding another deliberate mutilation of stock carried out in the interests of husbandry that is often likened to the tail docking of pigs. Widely practised in the poultry industry, beak clipping is believed by Webster (1994:111-112) to be another surgical procedure that is merely a means of controlling a problem rather than of solving it. In describing the situation that motivates stock keepers to carry out this operation he presented what many poultry keepers would recognise as an all too familiar scenario. An episode of mindless aggression breaks out in a group of birds. This takes the form of mutual feather pecking, which in turn results in the skin of one of the pecked birds starting to bleed. The attackers then become more aggressive and unless the bird is removed it is eventually killed and eaten by the other birds.

This behaviour is neither unnatural nor unique to large colony poultry systems. As Webster (1994:111) recognised: 'Hens are clearly motivated to establish 'a pecking order' or dominance hierarchy. This is an adaptive form of aggressive agonistic behaviour which should, in a satisfactory environment lead to a stable social community and so avoid further conflict.' In small groups of birds, however, the one that is attacked either has the opportunity to escape or has the chance of being rescued by its owner. In a large colony flock consisting of thousands of birds the likelihood of either happening is much reduced. As a consequence the stock keeper reverts to a solution that makes it difficult and unpleasant for the birds to behave in this way, that is by clipping the birds' beaks.

The dilemma again appears to be that of a cost/benefit equation. As FAWC (1991:24) stated, 'Unfortunately it is not known how to prevent outbreaks of cannibalism. We are faced therefore, with a most unfortunate dilemma: do we accept beak trimming as a preventative measure for the present or an increased risk of cannibalism?' The conclusion therefore is that 'The welfare problems caused by cannibalism are arguably greater than the welfare problems caused by beak trimming, provided that the operation is undertaken by a trained and competent operator.'

As Webster declared, this does not deal with the root cause of the problem, in that the size of the flock is too large either for the exercising of good stockmanship or for securing the welfare of individual birds. On both scores, therefore, I argue that the practice of beak trimming is not in accord with the principle of respect. Moreover, in removing part of the beak, even though it is only 'one-third of the beak, measured from the tip towards the entrance of the nostrils' (FAWC, 1991:24) the practice further offends the principle on two grounds. Firstly in that it denies respect to the essence of the bird, and secondly in that the bird experiences a form and degree of suffering that it would not have had to endure in the wild.

As regards the first, Webster (1994:111) described how the use of the beak is an essential part of a hen's behaviour pattern. 'For the hen, the beak is a major tool for investigating the environment so that they peck at all sorts of things, motivated not only by the desire to feed, or acquire nest material, but also by curiosity or simply mindless instinct.' In the barren and possibly oppressive environment of a large flock, birds might engage in feather pecking simply because there is little opportunity for them to peck anything else (Stevenson, 1994:12), although this is but one of a number of possible theories regarding the reason for this aggressive behaviour. It is obvious that in preventing the birds from pecking each other, beak trimming also effectively prevents them from pecking anything else. An important part of the behaviour pattern in terms of their essence as a chicken, is consequently denied. As a result I suggest that this practice is not in accord with the principle of respect and is therefore unacceptable.

Some argue that the application of a cost/benefit equation allows the practice to be deemed acceptable, as the effects of cannibalism are worse than the pain inflicted by beak clipping. According to Fraser and Broom (1997:383), however, the degree of pain experienced by a de-beaked bird can be extreme and long lasting. 'It has been stated that the part of the bill which is removed is merely horny material with no nerve supply so that birds feel no pain at cutting or later. This is certainly wrong... the chicken's bill has many Herbst's and Merkel's corpuscles, which are sensory receptors, and many free nerve endings in the part which is cut off during beak trimming. The removal of the end section of the bill causes a non-regenerable loss of touch and temperature sensitivity... scar tissue remains on the beak stump and neither dermal structures nor nerves regenerate into this scar tissue. The damaged nerves develop into neuromas which continued to grow for at least 10 weeks after the operation. In general, in animals neuromas are painful so it is likely that the chickens suffer pain for a prolonged period after beak-trimming.'

Following Rolston's conclusion (1988:79), that according to the principle of respect a farmed animal should not be caused more suffering than it would have experienced in the wild, I would argue therefore that it is difficult to justify the practice of beak trimming as a means of controlling feather pecking in a poultry flock (Sykes, 1994:217). The cost/benefit equation in this case does not work.

Whilst the problems of feather pecking and cannibalism are only too real in many poultry flocks, it is equally clear that there is no known reason why such an outbreak should suddenly occur. Sykes (1994:217) likened the sudden appearance of this aggressive behaviour to a disease, which 'spreads through a flock and then subsides.' Some have argued that it is a response to the over-stocked environment of a large colony flock (Stevenson, 1994:13), others that it is caused by 'vitamin or amino acid deficiencies' or 'the stimulus of the sight, smell, taste or blood' (Sykes, 1994:217). Sainsbury (1986:83) argued that the causes are multi-factoral and include 'an inherent tendency in certain strains, bullying, overcrowding, unevenness in the group, absence of or poor litter, excessive lighting, feed deficiencies or imbalances in quality or quantity, parasitism, poor growth, and deficiencies in ventilation or environmental control.'

Research into the causes of this aggressive and destructive behaviour has to be a priority for the poultry industry (FAWC, 1999:10). Only when the causes are established can a satisfactory solution be found and the practice of de-beaking become unnecessary - a practice that

Sainsbury (1986:83) argues has 'done nothing to remove the root cause' of the problem and which according to the principle of respect cannot be justified. As a result I would argue that those systems that require this surgical procedure cannot be accepted and that on moral grounds the producer should consider either adopting an alternative production system or leaving the sector completely. If either approach proves to be impractical then with Sainsbury I suggest that beak trimming must 'be looked upon as a transitory procedure pending an adjustment of the system to remove the danger'.

6.3   The impact of biotechnology

During the late 1990s a great deal of media attention focused on the influence of biotechnology on crop production, especially with regard to the modifying of the genetic structures of a variety of commercially grown plants, which became identified in common parlance as genetically modified organisms (GMO).

Defined by Kilpatrick (1997:30) as 'the application of understanding of living organisms and their components to create new processes and products' biotechnology has more recently 'become associated with molecular biological techniques which permit a single gene of one organism to be copied and incorporated into the DNA of another to create a GMO. Thus it is possible to introduce or eliminate specific traits in, for example, a crop plant, which otherwise possesses a wide range of desirable features, whilst avoiding the random shuffling of tens of thousands of genes through conventional breeding techniques' (Kilpatrick, 1997:30). Public concern at the possible introduction of these products into their diets, the description of such products as 'Frankenstein Foods', and the widespread anxiety over the possible impact that fields of genetically altered rape, maize and sugar beet may have on the environment have all stimulated opposition and protest ('The Week' 20.2.99:9).

Whilst there have been similar major biotechnological advances in livestock farming, in the main these have not created as much concern or protest. I believe that this has largely been due to the public's attention being focused on the medical benefits offered by animal biotechnology rather than on the technology's agricultural potential. This would seem to be confirmed by the case of the sheep called Tracey. The subject of a research programme at the Roslin Institute in Edinburgh during the 1980s into the production of a human protein in sheep's milk, she was cosseted and protected and received more attention than any of her farmyard cousins. The quality of the care she received, plus the potential benefits of the research for people suffering from emphysema, has meant that rather than provoking criticism the technology has met with widespread, if guarded support (SRT 16.11.99:2). Furthermore, I believe that this in turn has influenced the tenor of the whole debate regarding the relationship between biotechnology and animal production.

Nevertheless there have been developments in commercial livestock farming as a direct result of biotechnology that need to be examined in the light of the respect principle. Some have been beneficial to animal, producer and consumer, and can be supported by the principle of respect, whilst others have ignored the essence of the animal, have had a detrimental effect on its welfare and under the scrutiny of the respect principle can only be rejected.

Some of the concerns over the adverse effects of genetic manipulation on livestock were highlighted by Tudge (30.8.97:1) who described how 'muscle-bound animals are obese even before they have the bones to support their flesh. For this reason modern turkeys can hardly stand, and poultry handlers often break the bones of chickens just by lifting them from the cages.' Whilst this can often be true, I would argue that Tudge has failed to appreciate the fact that the problem is not to do with the technology being used but with the moral perceptions of the breeder.

When the essence of the animal is disregarded and the attitude of the breeder is one of disrespect towards that animal, then problems such as those outlined above are likely to result, regardless of the breeding practices employed.

Moreover it appears that Tudge was unable to appreciate the differences between breeding methods. Indeed, he seems to confuse the genetic manipulation of an animal's characteristics as the result of

237

a long-term conventional breeding programme based on careful selection of sire and dam, with the engineering of an animal's genetic make-up through the altering or inserting of a foreign gene by a human agent. Either process, when irresponsibly used, can result in the sort of cruel consequences he describes.

The first of these processes has been in use ever since man first domesticated farm animals (Hagedoorn, 1950:52), and many livestock breeders have long recognised the responsibility that this places on them in terms of the management and care of their stock. Indeed, criticisms similar to those made by Tudge were directed at those nineteenth century livestock producers who by selective breeding managed to rear animals whose appearance and characteristics offended many agricultural commentators of the time.

Discussing the condition and type of animals being bred for the Victorian show ring by wealthy and often ill-advised owners, Ritvo (1987:75) commented that *'Punch* frequently remarked on the shortness of breath of the prizewinners and the difficulty they found in moving. Most competitors were delivered to their show pens in special wagons, some had to be carried where they could not be wheeled. (Pigs were even worse than cattle in this regard, some highly admired specimens were so obese that bulging flesh obscured both their legs and their foreheads).' Ritvo suggested that such monstrosities were often the result of the breeder's desire to emphasise his own ability to manipulate an animal's physique, rather than to produce healthy and efficient livestock for commercial farming.

This suggestion was also made by many of the agricultural writers of the day who were quick to comment on the irresponsible approach of such breeders. The judges at the Smithfield Club's Christmas Show in 1821, for example, who continued to give preference to such stock, were said to be 'misguided' and the animals they awarded prizes to were described as 'gross, coarse-boned and excessively fattened' (Ritvo 1987:76). It was also in 1821 that the Board of Agriculture held its one and only agricultural show, only to find that it was disrupted by protesters who complained 'that although the purpose of the show was to encourage the production of good breeding stock, the premium had been awarded to a bull "which was too fat to serve"' (Ritvo, 1987:76).

From incidents such as these it is clear that responsible breeders of the period were unwilling to accept the imposition of breeding practices and objectives which they considered offensive and impractical because of the type of animals being produced and the suffering inflicted upon those animals.

Representing a later generation, Fraser (1954:137) similarly criticised the work of some of the stock breeders of his own era whose aims he believed were 'often vague, sometimes confused, occasionally completely wrong.' His main concern was that, as in the previous century, animals were again being bred for their looks and for presentation in the show ring, with little regard being paid to qualities required for life in the real world in which they would often have to face harsh landscapes and adverse weather (Fraser, 1954:139). In a similar vein Hagedoorn (1950:248) described how the influence of the show ring and the fashions it stimulated regarding a breed's particular attributes 'almost always counteracts progress in the breeding of economically valuable animals.' Animals need to be bred for the environment in which they are going to live and it is the breed characteristics that will enable them to thrive and be productive in that environment that need to be, and should be, encouraged.

This argument, which is propounded by both Fraser and Hagedoorn, takes due account of the animal's needs and nature, and thus of its essence, and emphasises the breeder's responsibility in terms of management and stockmanship. In other words, it pays due regard to the principle of respect and all that it entails in terms of practical stock keeping. As both authors emphasise, however, along with the agricultural commentators of the nineteenth century, when these criteria are ignored, the traditional/conventional methods of breeding livestock can result in the irresponsible production of animals whose very nature condemns them to a lifetime of suffering. As this was true in the nineteenth century and in the mid years of the twentieth century, so too it remains true as we enter into the twenty first century. As Tudge recognised, and as Donald and Ann Bruce (1998:110) have commented, 'the ethics of such 'traditional' methods are increasingly called into question.'

The focus of the Bruces' concern, however, was wider than the world of conventional livestock breeding. Their attention was firmly focused on those more recent developments in the world of biotechnology, which have been outlined above, whereby the required type of animal with particular attributes can be produced through the altering of that animal's genetic structure.

The rapidity of the developments in this area of biotechnology are illustrated in Table 12, but can also be seen in the major advances over little more than a decade that have taken place at the Roslin Institute in Edinburgh. Reference has already been made to the Institute's work in the 1980s, when a human gene was introduced into a sheep called Tracey. This enabled the ewe to produce the protein alpha-1-antitrypsin in her mammary gland, which could then be used to treat human emphysema sufferers. The development of this technology enabled the Institute in 1997 to produce Dolly, the first cloned sheep and the result of 'the technique of nuclear transfer to produce genetic farm animals from a cell culture, something which had not hitherto been possible' (Bruce and Bruce 1998:x).

As with their work with Tracey, the research project involving Dolly focused on the possible application of the technology to human medical needs. The ethical debate consequently tended to focus on the implications of using animals in this way and for these ends (BBSRC, 1996:14). At the same time, however, this emerging technology opened up new possibilities for the livestock producer, which have raised their own specific ethical concerns. One technique, for example, that could possibly be exploited in the animal industry, would be the insertion of genes for growth hormones into fattening stock, which would lead to a more efficient production of meat (Bruce and Bruce, 1998:116). So far, however, the results have been less than encouraging and have had some unfortunate and distressing consequences for the animals concerned. The 'Beltsville transgenic pig' for example, which was born 'arthritic, partially blind and with ulcers' (Reiss 1998:162) served to illustrate how this technology can condemn animals to a painful existence and death.

240

Table 12 , **Approximate time schedules for the introduction of animal biotechnologies**
(Source: Britton, 1990:55)

Biotechnoloy	Products	Introduction
Genetic engineering	Production of pharmaceuticals	1988
	Control of infectious diseases	1988
	Improvements in animal production	2000
Reproduction	Fertility improvement	1990
	Multiple ovulation and embryo transfer (MOET)	1990
	MOET and genetic engineering	2000
Regulation of growth and development	Hormone control of growth production	1990
	Immunological control	1995
	Genetic control	2000
Nutrition and feeding	Rumen microbiology and digestive physiology	1995
	Genetic engineering improvement	2000
	Grass and forage utilization	1988
Disease control	Diagnosis	1988
	New vaccines	1990
	Immunology	1988
	Selection for disease resistance	1995
	Genetic engineering resistance	2000

In this particular instance I suggest that this situation has arisen because the respect principle has been ignored, and the research objectives have solely been driven by the prospect of potential increased production and financial gain, with little regard being paid to the animal's welfare. Its essence, in terms of its needs and its well-being, has been ignored. It is not necessarily the technology that is wrong, but the way in which that technology has been applied and the use to which it has been put, in that both have taken place without any thought for the nature and needs of that animal.

On the other hand, some features of the work in this area 'are likely to be neutral or positive for welfare' (Bruce and Bruce, 1998:117). As an example Bruce and Bruce cite 'Herman the transgenic bull'. In this case the technology is aimed at producing milk from Herman's progeny containing the human protein lactoferrin, which makes it easier for babies and patients receiving antibiotics to digest. By nature and in normal agricultural circumstances Herman would serve cows, and any female offspring would produce milk. There are clear benefits for humanity, with no welfare problems for the animal; therefore the welfare implications following the application of biotechnology are in this instance neutral.

Similarly, North (1990:53) described the potential for producing 'transgenic animals with improved disease resistance'; an improvement which once achieved will be inherited by the next generation and become a permanent characteristic in those animals. If successful, this could prove to be an application of biotechnology that will lead to major health and therefore welfare benefits for livestock. It has to be said, however, that the Bruces (1998:117) are more pessimistic about such an eventuality. Believing that the complex nature of disease organisms makes their successful genetic alteration unlikely in the foreseeable future, they also argue that their ability to mutate would mean that they could quickly become infective once more.

What is certain is that, just as the field of animal biotechnology is still in its infancy, so too the ethical debate into its application is only just beginning. As the technology develops, questions and issues of an ethical nature will continue to arise and some would indeed argue that the commercial agricultural application of that technology should wait until some of the practical and ethical issues have been resolved. With regard to the cloning of farm animals, for example, the Farm Animal Welfare Council (FAWC, 1998:16) 'have found no aspect which we felt was intrinsically objectionable to the extent that it might be considered something that should not be done at all.' They proceeded to state, however, that 'this technology might, in the absence of suitable controls, result in significant insult to the nature and welfare of the animals involved' (FAWC, 1998:16). As a result of this and allied concerns they have called for 'a moratorium on nuclear transfer cloning in commercial agriculture while further investigation is made of welfare problems and uncertainties over oversized offspring, perinatal and birth problems, and aged DNA' (SRT 13.1.99:2).

Others have criticised the way in which some unscrupulous producers may use positive welfare benefits from biotechnology in a way that is detrimental for the animals being produced. Representing some of these concerns with regard to improved animal health as a result of biotechnological advances, Wrathall (April 1996:14) commented that 'animals better able to overcome exposure to disease might also tolerate more intensive or lower quality housing conditions. Increased resistance could lead to more animals being kept in inadequate conditions and result in various associated physical and mental welfare problems. The same concerns apply to altering genetic make-up to bestow tolerance to extreme temperatures.' In response I would argue that, in this instance, the argument should not be directed against the technology, but its application. The irresponsible and unethical actions of the producer are to be censured and the means to do this can be found through legislation.

Further criticism has come from those who believe that the technology is 'unnatural' and that the 'scientists are playing God' (Kilpatrick, 1997:30). It has to be said, however, that this has often been argued without the question being raised as to how 'natural' is to be defined or without entering into a debate on the role and activity of God in this context (Reiss, 1998:161). Their comment raises the interesting and often vexed theological questions as to whether God in this instance might be working through those scientists and whether they might be working in partnership with him rather than usurping his position and authority. In a similar vein 'some people will find gene transfer in animals unacceptable because they oppose all scientific intervention with animals... Other people may feel that it is intrinsically wrong to use laboratory techniques to move genes across any species barrier. For some, the only important distinction may be between human and

243

non-human species' (BBSRC, 1996:3). Such concerns will inevitably continue to pursue the application of biotechnology in the world of animal agriculture. I submit that the debate should include reference to the respect principle as a means of assessing the acceptability or otherwise of the practices under debate.

It is certain that with gathering momentum biotechnological research will continue to provide the livestock producer with new processes and practices that will have both positive and negative implications for the animals. As Foresight (1998:8) have reported 'embryo transfer is now being used in pigs and cattle to transmit high value genetics but it is not yet economic for commercial production. Semen sexing to increase the efficiency of livestock production is still under development. Ultimately, the production and storage of sexed, cloned embryos could give precise control of livestock production.' Such developments could be of untold benefit to the farmer struggling in the current economic crisis and, as it is likely that the successful commercial application of this technology is now on the horizon, I suggest that they are a source of hope for the industry which has to be pursued, rather than unfortunate developments to be rejected.

It is crucial that in both the research and subsequent practical application of biotechnology, the principle of respect for the animal's essence must be upheld. Lawrence *et al.* (1999:9) followed a similar line in arguing that there needs to 'be a greater emphasis on the welfare impact of genetics than in the past.' They also stated that 'the likelihood of some irreconcilable differences between production and welfare demands for breeding and the advent of new biotechnologies suggests the need for a mandatory ethical appraisal system for livestock production similar to that used in animal experimentation.' Unfortunately they offer no guidelines as to how this might function, but their suggestion regarding such a mandatory system is one that could well provide the framework for informed critical assessment of the application of biotechnology in agriculture. Regardless of whether such an initiative might be taken I would still argue that, providing safeguards such as those outlined above are in place, then there is no moral reason for rejecting a technology which can be beneficial to both animal and owner.

CHAPTER 7

Factors Affecting the Application of the Respect Principle

The principle of respect requires that the livestock farmer respects the essence of his animals through the exercising of responsible management and high quality stockmanship. If this principle is accepted as the ethical basis for modern livestock systems, then it follows that those influences which are likely to affect the adoption of the principle for good or ill need to be identified and understood. Only when this process has taken place can a strategy for either reducing or intensifying their impact can be produced, leading to a greater acceptance and application of the principle in the livestock industry.

7.1   The Cost of Respect

Many of the livestock systems that have been discussed above, and that are now criticised by a number of today's consumers on welfare grounds, were originally developed in response to the demand from a previous generation of consumers for cheaper food. Indeed, there are some grounds for arguing that the animal products that come in for most criticism (i.e. pig meat and poultry) still continue to grow in popularity with consumers. Table 13, for example, indicates that pig meat and poultry both increased in popularity between 1997 and 1998.

As Webster (1994:129) commented: 'the systems of meat production that attract most criticism on welfare grounds are those that attract most custom in the supermarket... They are cheaper.'

245

Table 13   **Household consumption of pig and poultry meats
in the UK**
(From MLC, 1999:60)

| | Consumption ('000 tonnes carcase weight equivalent) | |
	1997	1998
Pork	835	861
Bacon	459	473
Poultry meat	1549	1636

There is some evidence to suggest, however, that the diet and therefore the food purchasing habits of the British consumer are changing, especially with regard to animal products. As has already been illustrated, most citizens in an increasingly affluent society such as Britain have the luxury of being able to assess and criticise the way in which their food is produced with very little fear of going hungry as a result of those criticisms (Bennett and Larson, 1996:224).

As regards animal products, this has been reflected in the growing concern amongst consumers that the demand for an ever-cheaper product has been at the expense of animal welfare (McInerney 1991:23). As the RSPCA (undated) stated in describing the public's response to 'Freedom Foods', 'Increasing numbers of people are actively concerned about the issue of farm animal welfare and the Freedom Food scheme is responding to the growing demand from the consumer, for clearly labelled, welfare-friendly food. Many people are buying Freedom Food labelled products – and their purchases are making a positive difference to farm animal welfare.'

Some people suggest that the sense of disquiet concerning certain livestock production methods do not always influence consumer spending. This was noted in an article in 'The Economist' (April 1996:93) in which the comment was made that 'increased numbers of people in rich countries have doubts about modern farming methods.

But few, so far, are willing to give up the cheap food these methods have brought.' Twenty years prior to this Beresford (1975:175) came to a similar conclusion when he commented that 'we are not only sentimentally predisposed in favour of old-fashioned practices; we are increasingly worried about new practices that foul the nest. At the same time, with an inconsistency that springs from ignorance about how food is produced we are most of us unprepared to pay the surcharge that old-fashioned practices entail.'

Having already argued that many old-fashioned practices were less welfare friendly for the animal than their modern counterpart it would be difficult to support a general return to outdated and often morally unacceptable farming systems that do not conform to the ethical criteria already established. Moreover, there is sufficient evidence to suggest that the assumption that the public will refuse to pay for welfare friendly food is incorrect.

Bennett (1996:320) has recorded the evidence of a survey of 2000 people in the UK in which 62% of the respondents reported that 'they had altered their purchasing behaviour because of their (welfare) concerns (most commonly by purchasing free-range rather than cage eggs).' The survey's findings also indicated that 86% of the respondents supported the banning of battery cages even if it meant that they would have to pay extra for eggs. It is perhaps worth noting, however, that the willingness to pay varied greatly according to the extra costs involved (see Table 14).

In reporting the survey's findings Bennett noted that the final results cannot prove one way or another whether those surveyed would actually be willing to pay the additional cost of buying eggs from alternative systems to battery cages should the situation arise. Indeed, he suggests that there are a number of reasons why people may make a biased response to the questions asked, not the least being their desire 'to make a statement by... inflating their stated willingness to pay' (Bennett, 1996:321).

Despite Bennett's degree of cynicism, his research findings are very similar to those recorded in a survey of 2000 consumers conducted for the Royal Society for the Prevention of Cruelty to Animals (RSPCA)

247

in 1997. Their results indicated that 69% of shoppers were 'willing to pay more for a product which comes from a humanely reared animal. Of these 42% would be prepared to pay 50p or more, on a pack of meat costing £2.00 (RSPCA, 1997:2).

Table 14　　　**People's willingness to pay to support legislation to ban battery cages** (increase in current egg price in pence per dozen eggs) (Source: Bennett, 1996:321)

Willingness to pay	% of respondents
0	14
>0 to 20	10
>20 to 40	32
>40 to 60	18
>60 to 80	11
>80 to 100	7
>100 to 120	6
>120	3

- Numbers do not add up to 100% because of rounding

From figures such as these it seems that the majority of consumers are willing to pay extra for animal products that have originated from rearing systems which are welfare friendly, and therefore according to the criteria previously established, also ethically acceptable. This begs the question as to why, if this is the case, the general public still purchase products such as imported pig meat from systems that would be considered cruel and illegal in the UK.

It may well be that Bennett's cynicism regarding the link between expressed opinion and actual food purchasing habits is well placed, although I suggest that many consumers have simply been misled by the often vague labelling of food products regarding matters such as country of origin and method of production. Egg cartons carrying pictures of hens running free in a farmyard for example, and bearing the legend 'Farm fresh eggs' can hide the fact that those eggs may well have been laid by caged hens.

Moreover, most people in the UK are now so far removed from the world of farming that they have little knowledge or understanding of how their food is produced or of what the words on food labels really mean (Richardson, 'Farmers Weekly' 26.5.00:99). Whilst they may feel strongly about animal welfare issues, their ignorance means that they are often unable to relate those feelings to the food they buy in the supermarket. I believe that this is one of the main issues with which livestock producers have to deal. Before consumers can make informed judgements about the animal products they buy and the conditions under which those animals were reared, they first need to be educated about the origins of their food, how it was produced and what the terminology on the packaging labels actually means.

Nevertheless there is still the question as to whether the products from the more intensive systems are in reality cheaper than the alternatives. A comparison of prices at a supermarket checkout would suggest that the answer is obvious (Table 15).

Table 15   **Comparative retail price of eggs from different production systems** (Tesco Stores, Warwick, 10.3.2000) **and price paid by the packer to the producer** (NFU, July 2000:14)

	Retail (pence per egg)	**Price received by producer** (pence per egg)
Free range	1.60	0.78
Barn/Colony	1.39	0.60
Caged	0.76	0.29

Taking eggs from hens kept in battery cages as an example, as Table 15 illustrates, the price to the consumer is considerably less than that being charged for comparable eggs from free-range systems.

Despite this incontrovertible evidence regarding the comparable cheapness of eggs from intensive flock systems both at the point of sale to the packer and the consumer, Stevenson (1997:4) has argued that the price differential between caged and free range eggs does not reflect the actual differences in production costs. He has stated that 'free-range eggs... involve only slightly greater production costs than battery eggs. Stevenson supported his argument by comparing the actual costs of different production systems, although regarding his comparable egg production figures (Table 16) one could argue that his definition of 'slightly greater production costs' depends on the size of the units in question.

Table 16   **Comparison of production costs as between different egg production systems** (Cited by Stevenson 1997:6)

(A: as per survey by Roberts and Farrer, 1993)

	Battery	**Free-range**	**Perchery**
Total costs of bird per year (£)	10.10	12.80	11.57
Number of eggs per bird per year	278	254	255
Cost of producing one egg (p)	3.63	5.04	4.54

(B: as per NFU, 1997)

	Battery	**Free-range**	**Perchery**
Total costs per bird per year (£)	11.58	14.75	13.21
Number of eggs per bird per year	279	266	271
Cost of producing one egg (p)	4.16	5.56	4.88

Note: NFU (1997) gives total costs and egg yield for a 55-week laying period, which Stevenson converted into per annum figures.

On a farm with 10,000 fowls, using Roberts and Farrar's results, the additional cost per annum for a free-range producer would amount to £27,000 more than those of a battery unit of the same size. Again using the results of Robert and Farrar's research the total costs of that free-range system would amount to £128,000. On this basis I would suggest that Stevenson's description of this additional cost as insignificant is to say the least inaccurate!

In contrast to Stevenson's assertion that the cost differential between what he believes to be cruel and welfare friendly production systems is insignificant, Varley (undated:3) suggested that for pig producers the conversion to welfare friendly 'semi-intensive straw-based systems increase housing costs by about 20% or more and probably are associated with other increased costs such as labour.' In a similar vein an article in 'NFU Business' (August 1999:16) argued that the total cost to UK egg producers following the banning of battery cages under the Laying Hen Welfare Directive would amount to £172.45m. Another article in 'The Farmers Weekly' (June 1998:5), on the increase in free range broiler production in response to consumer welfare concerns, stated that the producer needs to earn a welfare premium that will 'compensate for a finishing time which is nearly double that for intensively reared birds.' Webster (1994:261) simply stated that 'the reason why the battery cage has become so dominant is that no other system can compete economically as a mass producer of eggs.'

It would seem from the above statements that Stevenson's conviction regarding insignificant differences in comparable costs between different production systems is misplaced. There are other factors, however, which whilst they do not appear above still need to be taken into account. One is the fact that products from livestock rearing systems that the consumer believes to be welfare friendly can command a healthy price premium that may well compensate for any additional production costs ('Farmers Weekly' June 1998:5). Varley (undated:2) suggested that the high welfare standard required of British pig producers provide them with 'an enormous marketing advantage', although surveying the state of the UK pig industry at the start of the 21st Century one wonders at the evidence as well as the conviction behind this argument.

As regards egg production, however, McInerney (1991:24) recorded how 'those who possess a clear preference (and can afford it) are ready to pay a premium of over 40% in order to consume eggs from free-range rather than battery hens.' Again, however, it has to be acknowledged that McInerney also appreciated how fickle this preferential market can be. 'What at one time might be an appreciable price premium can rapidly fall to marginal or uneconomic levels if sufficient (i.e. too many) farmers switch into supplying these products for what at present remains a minority market' (McInerney, 1991:25).

Similarly, if the consumer ceases to make chicken welfare a matter of priority and turns to purchasing the cheaper product, free-range production of eggs becomes uneconomic and unsustainable as a business enterprise. This possibility takes on added meaning when one remembers that the current interest in animal welfare issues is a comparatively recent phenomenon which has been linked to the growing affluence of an urbanised western society. One wonders therefore if a change in economic circumstance might well lead to a lessening of sensitivity and thus of interest in animal welfare matters (Bennett, 1995:55).

Whilst there is clearly some uncertainty, certainly in the long-term, over the direct economic benefits of those livestock production systems the consumer believes to be welfare friendly, there are other non-financial, non-farming and yet possibly enduring benefits that have to be taken into account. A Conference on the theme 'Farm Animals - It pays to be humane' identified two of these as benefit to the consumer (McInerney, 1991:19) and benefit to the stockperson (Seabrook, 1991:60). Regarding the first McInerney described how 'the benefits from improvements in animal welfare are manifested in the way we, as a society, feel better off as a result... primarily in the way we feel less disquiet about how livestock are exploited in our favour, in feeling more comfortable that we are not imposing what we think are unacceptable costs on animals.' He then concludes that as 'these feelings are something we value, they represent benefits every bit as real and valid as those we get from things with an obvious financial counterpart, such as things we buy.'

One might argue that the overall value of such an intangible benefit is not only difficult to assess, but is of little help to the livestock producer caught up in the current economic decimation of the UK animal industry. In contrast, Seabrook's view on the benefits associated with welfare-friendly systems promises direct economic benefits to a farm business. In addition I believe that they provide a further degree of encouragement to the depressed producer by enhancing his feelings of self-esteem through improved job satisfaction.

Seabrook's argument (1991:59-70) is that good stockmanship, which as has been recognised is a fundamental feature of the principle of respect, has a marked effect on an animal's productivity. In pigs that received what Seabrook describes as 'pleasant handling' there was not only clear evidence of better growth rate, but also more piglets per litter, plus a better piglet mortality ratio than that found in systems in which the animals received 'aversive treatment'. This is clearly indicated in Table 17. Similar evidence emerged from studies of the productivity of dairy cows subject to contrasting stockperson attitude and behaviour. Those, for example, that experienced pleasant handling were found to be higher-yielding than those subject to aversive treatment.

Table 17   **The effect of different handling treatments on farrowing sows**
(Source: Seabrook, 1991:65)

| | Handling treatment | |
	Pleasant	Aversive
Piglets born alive per farrowing	10.1	9.3
Piglet mortality at three weeks of age (%)	11.1	15.2

From the above it would appear that, whilst the implementation of welfare-friendly practices in livestock production might well impose additional costs on the livestock producer, such as in the extra amount of time that the stockperson spends with their animals or in the extra staff time required in the management of straw-based housing systems, there are certain benefits to both the consumer and the producer, which must also be taken into account. Furthermore, it could be argued that a cost/benefit comparison between different livestock rearing systems cannot be analysed fully until all costs are included in the equation. Indeed there is a growing body of evidence to suggest that much of the overall cost of producing food in many intensive systems is actually hidden and that the consumer is unknowingly paying extra for that food in a variety of indirect ways through rates and taxes.

Described by agricultural economists as 'negative externalities' Stevenson (1997:15) identified such costs as 'environmental pollution... an increased incidence of food-borne disease... and the poor welfare experienced by intensively reared animals.' The continued increasing scale of these costs which are directly attributable to intensive livestock systems has been highlighted by Pretty (1998:48), who commented that the price society has had to pay can be measured in terms of the environmental and social problems which he suggests these systems have created. In examining the environmental impact of intensive methods of livestock production, for example, he concluded that 'the atmospheric environment is contaminated by methane, nitrous oxide and ammonia derived from livestock, their manures and fertilisers. Despite considerable research, advocacy and political change over the years, these costs to national economies are still growing' (Pretty, 1998:48).

To a certain extent it is possible to put a monetary figure to some of these costs, as is illustrated by the situation regarding increased levels of nitrates in drinking water. Mainly caused by the application of inorganic fertilisers and animal wastes to farmland, it is known that resolving the problem costs £24 million per year (Pretty, 1998:50). On the health front Stevenson (1997:20) noted that 'it is generally accepted that BSE has been caused by the highly intensive practice of feeding to cattle the remains of sheep or cattle which proved to be infected.' Whilst some might dispute Stevenson's conclusion regarding the reason for the outbreak in UK cattle, the costs associated with the disease cannot be denied. If a definite causal link is established between BSE in cattle and vCJD in humans then the most tragic cost has been in terms of the lives that have been lost as a result of the disease. There has also been a tremendous fiscal expense for the nation amounting to billions of pounds (Stevenson, 1998:20) as well as a cost that is impossible to determine in terms of the ongoing impact of the disease on British livestock production. For example, it is unlikely that the expenses which have resulted from the imposition of a vast bureaucracy set up to administer and monitor animal traceability, along with the associated increase in charges at abattoirs for carcase inspection, will ever be fully assessed.

There are other costs attached to livestock production systems that are even more difficult to quantify. Pretty (1998:63) suggested, for example, that the demands of current methods of animal husbandry have meant that 'draining and fertilisers have replaced floristically rich meadows with ryegrass monocultures.' He believes that this in turn has had an adverse effect on bird life, removing the source of food and habitat on which many species were dependent. In terms of the pleasure and recreation that a rural landscape and the sight and song of birds provides for many people, he concludes that this loss is clearly a major cost attributable to current agricultural practices. Whilst it would be impossible to put a monetary value on that cost, he argues that it must still be taken into account when assessing the validity of the claim that modern production systems make for cheaper food.

From the above it is clear that a number of the concerns regarding current practices and methods in livestock husbandry extend far beyond matters of animal welfare. I suggest, however, that there are certain aspects of the principle of respect that are relevant to some of these issues. For example, stewardship, defined previously as one of the principle's two major components, is about the responsible use of resources in the implicit understanding that that responsibility is exercised not only towards the animals in one's care but also towards God and the rest of human society. On that basis, a system that is wasteful or destructive of the environment, which the religious believer sees as the handiwork of God and a gift for humanity to enjoy, and that is consequently destructive of human health and happiness, is morally unacceptable.

Further concerns regarding the environmental costs associated with modern livestock husbandry have concentrated on the global rather than the local implications of intensive rearing systems. It would seem that this wider assessment of cost has two main areas of concern. The first is to do with the feeding of cereals to livestock, which is the normal procedure in many modern production systems. This practice is believed by some to be an inefficient and wasteful way of utilising a valuable food commodity in a world in which the majority of the population goes hungry (Rifkin, 1992:180). The second is related in that it is argued that the growth and export of food crops from third world countries for animal feed in the developed world deprives people in those producing nations of the possibility of growing much needed food for themselves. This was commented on by Johnson (1996:59) who stated that 'around 40% of world grain production is fed to livestock, and animal fodder for export is an important cash crop for many of the poorest countries. This can divert resources that could better be used to feed the population at home, as well as often resulting in soil impoverishment and erosion.'

Whilst both are very real concerns, their resolution is far more complicated than the above statements might suggest. As Dower (1996) has illustrated, the problem of world hunger is not so much a matter of a global insufficiency of food, but is rather a consequence of many of its citizens being too poor to purchase the food that is available. 'The reason why the hungry are hungry is that they do not have access to food, because they do not grow it, do not have the economic power to acquire it, or are not given the money to buy it' (Dower, 1996:6). One should also add that in many famine situations such as Ethiopia, people go hungry because of war. Driven from their land, unable to purchase seed or to plant or harvest their crops, they starve.

Gower (1996:6) also commented that 'many studies attest to the claim that we have hunger in a world of plenty.' The food is there but beyond the economic reach of those who desperately need it. Whilst

Rifkin (1992:180) may criticise the feeding of cereals to livestock in the belief that it is wasteful, in market terms there is in fact a global surplus of cereals, which has resulted in a worldwide collapse of cereal prices and farmers across Europe being paid to take land out of production. On this basis it could be argued that, whilst there is such a surplus, it is as sensible to feed that surplus to animals as it is to pay farmers to take more land out of production as a means of regulating the amount that is grown. This would at least stimulate European rural employment and ensure that the land is kept in good heart.

This is not to dismiss the problems of world hunger, but to argue that the answer is to be found as much in the political and economic arena as it is in the agricultural sphere. Indeed, on this basis it could be argued that globally it makes more sense for the developed countries to increase their production of home-grown cereals for animal feeds instead of taking land out of production. This in turn would enable current exporters to focus their own cereal production on local needs (Korten, 1996:30). This argument, however, fails to take account of the fact that the present economic survival of those countries may well depend on their current trade in the raw material for conversion into animal feed.

Furthermore, I suggest that Rifkin's concerns (1992:200) regarding the disastrous ecological effects of over-grazing, especially in economically poor and environmentally delicate areas, which he blames on modern livestock production systems, are more a result of bad rather than modern farming practice. Rifkin (1992:200) argued that 'desertification (is) caused by the overgrazing of livestock; over cultivation of the land; deforestation; and improper irrigation techniques. Cattle production is a primary factor in all four cases of desertification.'

I suggest that this is far from being a modern phenomenon and that this generation is not alone in failing to manage grazing properly or in creating deserts in place of fertile fields. Freudenberger (1990:71) described, for example, how over the course of thousands of years overgrazing has destroyed the once fertile plains of Mesopotamia and the Nile valley, the Sinai Peninsula and the Trans-Jordan plateau, the highlands of Judea and Syria, and the lush forests of Lebanon. In fact rather than creating the problem I would argue that modern livestock systems that are based on the high standards of agricultural management required by the principle of respect could well help to resolve rather than cause or exacerbate it. Indeed, respect for an animal's essence will lead inexorably to a greater understanding of that animal's needs and how they should be met, especially with regard to their behavioural and nutritional requirements, and the impact that they are consequently going to have on their environment.

When this understanding is combined with those other features of modern livestock husbandry which are closely bound to responsible stock and land management, such as the application of biochemistry to the cultivation of fodder, the ongoing developments in soil science, and the employment of technology appropriate to the local situation, the net result may well be the establishing of production systems that are environmentally as well as welfare-friendly.

7.2   Consumer responsibility

One of the main issues that has frustrated and angered UK livestock producers at the turn of the millennium has been the matter of food labelling. Angered by the unilateral and illegal action of some nations in closing their markets to British beef as a direct consequence of BSE, and equally enraged by the flood of pig meat imports into Britain from systems deemed illegal in this country, UK farmers have then had to experience the frustration of seeing many such imports being placed on the supermarket shelf with a 'Produced in Britain' label. Some have argued that the legislation on labelling which allows this to happen and which permits processors, packers and retailers to obscure a product's country of origin by denoting only the country in which that product has been prepared for sale has been deliberately designed to confuse and mislead the consumer ('NFU Business' March 2000:13). Whether this is the case or not, what is certain is that the current food labelling regulations make consumer discrimination on welfare grounds, or for that matter any other ethical grounds, almost impossible. Equally, such unclear labelling removes any chance of the consumer monitoring how a joint of meat, an egg or a litre of milk has been produced.

The opportunity for the consumer to discriminate between one food product and another and to monitor how that food has been produced is an essential prerequisite to the demands by those same consumers for welfare-friendly products, especially when those demands are enforced by law. Indeed I believe that they constitute a moral obligation on the part of the consumer. Responsible and sustainable food production depends on an implicit social covenant between the consumer and the producer, as each has a stake in ensuring that there is an ongoing supply of food to which everyone has access (Pretty, 1998:11). On one side of this covenant the producer has a duty to the consumer to ensure that the food being sold is safe, is of good quality and meets the requirements of the buyer. On the other side, the consumer has an equal duty to the producer to ensure that there is no distortion of the food market by unscrupulous retailers, badly-framed legislation or continued public ignorance regarding the way in which that food is produced.

The need for such a covenant between grower and consumer has been recognised by the National Farmers Union (NFU), which stated in its call for a national long-term strategy for agriculture (NFU, mid-March 2000:3) that 'the Government must let farmers know what is required of them - and what farmers, in return, can expect from society.' This concept of partnership was later formalised in a proposal, which were presented to the Prime Minister by the NFU for 'a contract between British farmers and society' (NFU, 28th March 2000). I submit that the recognition of and commitment to this partnership is essential to the future well-being of agriculture, and that if it is ignored or weakened on either side then the potential for a sustainable food production system is similarly diminished.

This is Pretty's argument (1998:11). Having stated the obvious in describing everyone as a 'stakeholder' in the production of food, he then proceeds to identify eight groups to which those various stakeholders belong, ranging from input suppliers, farmers and rural communities, through to manufacturers, consumers, countryside visitors, environmental bodies and government. 'As connections are weakened, understanding quickly gives way to distrust and suspicion' between the various groups. This in turn leads to the need for a renewed 'collective responsibility' allied to a greater mutual trust, a partnership which I have already represented in terms of a covenant relationship. Pretty (1998:21) illustrated the practical outcomes of this trust by citing a variety of different schemes of which 'community-supported agriculture (CSA)' is possibly both representative and symbolic of all the others. 'The basic model is simple: consumers provide support for growers by agreeing to pay for a share of the total produce, and growers provide a weekly share of food of a guaranteed quality and quantity. CSAs help to reconnect people to farming, and farmers to their customers. Members know where their food has come from, and farmers receive payment at the beginning of the season rather than when the harvest is in. In this way the community shares the risks and responsibilities of farming' (Pretty, 1998:22).

Whilst this model clearly creates difficulties for cereal farmers, in that few consumers are going to be interested in purchasing raw grain, it does have positive implications for those livestock producers who are not as distanced from their customers by numerous and necessary stages of food processing. The sale of eggs, meat and some dairy products, for example, as with certain horticultural produce, can be done at a relatively local and personal level and thus a close link and therefore partnership can be established between purchaser and producer. I would argue, however, that such a local application of Pretty's vision is far too restrictive. Indeed, I believe that his vision must be extended from the local to the global, to cover the whole area of food production and marketing.

When society becomes aware of the implicit covenant that needs to exist between producer and consumer, and of the commensurate fact that we all have a stake in how our food is grown and produced, it is then that a responsible and sustainable strategy for farming can be established. This is not a matter of purely local interest or concern, but one which has global dimensions. When, as Pretty suggests, 'the community shares the risks and responsibilities of farming' either through political support or consumer action, a mutual trust and dependency can be established which can then become the foundation for a stable, efficient and economically viable agricultural industry that is above all ethically acceptable (Pretty, 1998:302). I believe that this can be achieved at a national and even an international level, rather than locally, and that that achievement is essential for the future well being of our world, human and non-human. Responsibility for upholding the ethical principles on which agricultural systems and methods are based rests with all of society and not just those directly involved in farming. To exercise this responsibility, however, consumers have to know how and where their food is produced.

This takes us back to the earlier concern regarding food labelling. Pretty (1998:24) argued that honest labelling is essential to 'increase consumer confidence about both the source and quality of food.' I would add that this would also enable the consumer to support and encourage those producers who are providing food of a high quality that is grown and reared in systems that society deems ethically acceptable. As I have argued that both intensive and extensive livestock production systems should be based on the principle of respect, so I would also argue that the wider adoption of the principle, and all that it entails, could be encouraged by consumers being able to support such systems by choosing to purchase products from them. Such choices are made possible when the consumer is confronted by clearly produced and honestly presented food labelling.

Indeed, it is largely for this reason that farming organisations in Britain have pressed the government to support them at two levels with regard to the shelf description of food products ('NFU Business' March 2000:10). The demand is for honest unequivocal labelling of a product's country of origin allied to the adoption of an assurance mark for high quality produce. In February 2000 Tesco responded by committing themselves to 'clearer country-of-origin labelling' and a 'standardising of quality assurance' ('NFU Business' March 2000:10). Following Tesco's commitment pressure is now being exerted on other retailers to follow their example.

The value of a widely accepted standard for quality assurance, which has been highlighted by Tesco, has long been recognised by producers and their representative organisations. Indeed, the rapid growth in the number of quality assurance schemes during the 1990s was accompanied by increased concern that the proliferation of schemes was more to do with retailers trying to score marketing advantages

261

over each other rather than a desire to establish a national standard for high quality food ('British Farmer and Grower' April 1997:11).

There is evidence of a further degree of cynicism from many livestock farmers who believed that some of the required measures regarding animal welfare may well have been inspired by ignorance and sentiment rather than established scientific evidence (Wright, Sept. 1999:42). Such cynicism was felt to be well-founded amongst pig producers when the RSPCA's Freedom Foods assurance scheme outlawed the use of farrowing crates in 1999 and as a consequence pre-empted research that had just been embarked upon by Cambac JMA Research into this very area. Whilst one can accept that the sight of sows restrained in a farrowing stall may be distressing to those with little or no knowledge of pigs, there is no excuse for a leading animal welfare organisation to deliberately ignore well-established welfare arguments which have long supported the use of farrowing crates (Wright, Sept.1999:42). To then effectively predict the outcome of research into the subject must be deemed totally unacceptable, not only because of the additional financial pressures that have consequently been placed on farmers who have to redesign their rearing systems, but also because the move may well have exacerbated rather than relieved animal suffering, especially for piglets (Wright, Sept. 1999:42).

Tesco's call for a nationally accepted standard for farm assurance schemes has possibly found an answer in the NFU's plans to adopt a British kite mark for UK agricultural produce. With the objective of 'creating and promoting to consumers a single logo or kite mark that can be used on all British farm assured food' the scheme is 'underpinned by the requirement for products bearing the mark to be farm assured, from schemes with independent verification and standards that ensure food safety, responsible environmental management and good animal welfare' (NFU, undated).

Unfortunately, however, whilst the initiative deals with a number of the concerns already mentioned with regard to food labelling, it still leaves the matter of a national standard for both animal welfare and food quality unresolved. Until such time as the government's Food Standards Agency may be able to establish such a national standard for all farm produce, however, the nearest approximation to such a standard is likely to be British Farm Standard food label, which was launched by the National Farmers Union (NFU) in April 2000. Covering a range of produce, its standards will be monitored and verified by the various currently existing assurance scheme operators. The policing of the scheme will be carried out by The British Farm Standards Council, which consists of representatives from all 'the participating British farm assurance schemes plus other supply chain stakeholders and independent members' and which will also grant licenses for the use of the logo (NFU Business, April 2000:2). Whilst the initiative is commendable it still fails, unfortunately, to provide the consumer with a universally accepted welfare standard by which they can compare one animal product with another.

The major omission of a single standard of animal welfare in the establishing of a nationally accepted assurance scheme may well be filled, however, by two other schemes - Farm Assured British Pigs (FABPIGS) and Farm Assured British Beef and Lamb (FABBL) both of which are based on the requirements of FAWC's 'Five Freedoms'. The reason for such optimism is that these two assurance schemes tend to be the benchmark by which others are measured or to which they are linked. For example, the Quality Standard Mark for Pork is linked to FABPIGS and Lincolnshire Quality Beef and Lamb to FABBL. In addition, as from April 1st 1999, FABBL standards incorporated those of another major player, Assured British Meat, a move which has been described as 'the first major step to delivering national unified farm standards' ('The Standard,' Spring '99:1).

One can only hope that this move will enable both producer and consumer to fulfil their mutual responsibilities and make the implicit covenant between them a living reality.

7.3 Agricultural research

As was suggested in Chapter 6, much of the ongoing ethical debate regarding the role and application of biotechnology in the world of commercial animal production will be informed and influenced by continued scientific research. As, for example, our understanding of

263

the genetic structure of disease organisms becomes clearer and our knowledge of embryo technology continues to develop, so the potential for practical on-farm application of this technology will become more realistic and the ethical debate more focused.

Until this happens, however, such concerns are likely to be of peripheral interest for most producers. Whilst a farmer may appreciate the potential benefits inherent in biotechnology, that technology will only become relevant practically and ethically when it is likely to enhance an enterprise's economic viability or provide other benefits, such as higher welfare standards, at a price he can afford.

There are areas of agricultural research, however, which I believe can and do fulfil these criteria. Not only are the results of such research likely to have implications for the economic viability of many livestock enterprises and the welfare of stock, but I would also argue that they encourage the wider adoption and application of the respect principle. Some of these areas were highlighted by FAWC (1993) in their report on the 'Priorities for Animal Welfare Research and Development'.

A later summary categorised these priorities under five headings: livestock breeding programmes, stocking densities, mutilations, disease and stockmanship, and technology transfer (FAWC, 1998:9). Certain of these areas have clear economic implications for the profitability of the UK livestock industry. Indeed, the disastrous effect that BSE has had on that industry since 1996, when the possible link with vCJD was first announced, combined with the continued cost to producer and public alike, serves to illustrate the economic importance of all research into animal health and welfare. This was reflected in the report published by MAFF (1996) on the subject of research priorities for the Ministry. Regarding animal health the authors stated that 'the economic well-being of the UK livestock industry depends on the maintenance of high health status which underpins exports of live meat animals and genetic material amounting to just under £1.25 billion in 1994' (MAFF, 1996:38).

Those comparatively prosperous days are now a distant memory for British livestock producers, simply because the warning came too late.

264

In March that year BSE decimated the national cattle herd and subsequently brought commercial animal production in the UK to its knees. It is to be hoped that the lesson that animal health has to be a major priority in any strategy for research involving livestock has now been well learnt, for the economic survival of the industry may well rest on its outcome. Without even taking BSE into account it is estimated that 'the cost of endemic infectious diseases to the UK livestock industry is 17% of the value of farm gate sales (about £1.5 billion per annum)' (Foresight, May 1998:8). Costs as great as this must surely emphasise the need to prioritise further research into animal health matters.

Work in this area of agricultural research has not been motivated solely by concerns for those producers whose profitability may well have been affected by the onslaught of disease in a herd or flock. Rather, in calling for further scientific study into such problems as lameness and mastitis in dairy cattle or foot rot in sheep, FAWC's priority (1999:10) has been the alleviation of any distress that the affected animals were likely to suffer. In this I suggest that the Council's motives express the very qualities of care and responsibility towards stock that are integral to the principle of respect.

In addition, as this research into the susceptibility of livestock to a variety of ailments and diseases is also likely to lead to a better understanding of the experiences and perceptions of the animals being studied, so in turn it should encourage a greater awareness of the nature, needs and character of those animals. As a consequence this should enable those responsible for the care of stock to appreciate more fully and respect their essence, and so further promote the respect principle in a practical agricultural environment.

On this basis it is not surprising that in prioritising this area of research FAWC also link it with the need for further study into the whole area of stockmanship (1998:9). Indeed, it would be difficult to consider the one without the other as the responsibility for much of the diagnosis, medication and care of sick animals rests with the stockperson (FAWC, 1999:10). Seabrook (1984:87) took this further and argued that the stockperson's ability and skills affect more than an animal's health. He stated that there is sufficient evidence to suggest that the standard of stockmanship in any unit can have a marked effect on an animal's performance. Whilst he appreciated 'that there is no scientific epistemology that will readily provide incontrovertible evidence to precisely quantify the role of the stockperson in influencing the performance of animals' (Seabrook, 1991:59) he was still prepared to argue that the evidence is conclusive enough for this deduction to be made. Such evidence is illustrated in Table 18.

Table 18   **Behaviour of stockperson in the milking parlour in relation to dairy herd milk yields in one-person units**
(Source: Seabrook, 1991:62)

Herd output	Behaviour of stockperson in milking parlour	
	(number of times per minute per cow)	
	Touching cows	Talking to cows
Higher yielding	2.1	4.1
Lower yielding	0.5	0.6

Seabrook's conclusion draws attention to what I believe is an indisputable link between the way in which the application of stockmanship skills can improve an animal's productive performance and also enhance its welfare. 'Research has shown that the behaviour of the stockperson can be a potential stressor with consequences for productivity and welfare. By adopting the correct empathetic behaviour the stockperson can have a crucial role in creating an environment for improved animal welfare' (1991:69). This is clearly an important insight into one of the most fundamental aspects of livestock husbandry.

To achieve wider recognition and acceptance of Seabrook's findings I suggest that much still needs to be done. English *et al.* (1992:16) commented that 'although the importance of stockmanship in relation to animal welfare and animal performance appears to be widely recognised, little or no action has been taken by the relevant authorities to encourage research into stockmanship.' This would seem to be confirmed by Seabrook and Wilkinson (1998) who, in their study of 'The Stockperson as a Resource on Dairy Farms', identified a number of key areas that required further research. These ranged from matters pertaining to practical husbandry through to issues concerning the effect of socio-economic changes on the stockperson (Seabrook and Wilkinson 1998:29).

Committed and caring stockmanship has previously been identified as one of the two criteria for assessing the application of the respect principle in any production system. As a result, when arguing for the wider adoption of the respect principle in the world of UK livestock production, one can only applaud any effort to promote the importance of stockmanship skills and qualities. This is exactly what Seabrook and Wilkinson (1998) and English *et al.* (1991) have done and what FAWC (1999:10) have advocated in recommending that there should be further research into 'the education, motivation and continued training of all those involved with livestock.'

Just as the value of the research programmes mentioned above can be appreciated in terms of both economic and welfare benefits, there are other areas of research, such as the study of livestock breeding, that can achieve similar results. Indeed, the history of conventional selective animal breeding, which has already been briefly touched on earlier in this chapter, is closely allied to the prosperity of British agriculture. This is particularly reflected in the work of the agricultural improvers of the eighteenth and nineteenth century, whose achievements have to be appreciated in economic as well as aesthetic terms. Whilst some enthusiasts spent great amounts of money producing animals that had no relevance for practical commercial farming, others such as Robert Bakewell and John Ellman concentrated on breeding stock that would be of benefit to the farmer (Webster, 1994:54). Concentrating on specific qualities that would improve the fertility or productivity of an animal, they carefully selected sire and dam and produced new stock that was more commercially and agriculturally viable than anything that had gone before.

This emphasis on productivity caused some irresponsible breeders to rear monstrosities. In the main, however, livestock producers are still enjoying the achievements of these improvers, especially as their work has become the reference point for all subsequent selective livestock breeding programmes (Fraser, 1954:83-87). The desire for greater productivity, which probably provided most of the motivation for men such as Bakewell and Ellman (Webster, 1994:55), still dominates much of the research into livestock breeding.

There have been changes, however, not the least being the advent of artificial insemination (AI). It is fair to say that the introduction of this particular technique to livestock farming has completely transformed the world of the commercial stockbreeder. Although its widespread use in the UK cattle herd only began in the early 1940s (Fream, 1973:522), AI has now become an integral part of breeding programmes throughout the livestock industry (Leaver, 1994:23). Its main impact has been within the dairy sector where it has enabled breeders to use bulls that have been far superior to anything that they might otherwise have been able to afford. This in turn has resulted in British dairy farmers achieving productivity figures that would have been unimaginable to a previous generation. 'Whilst herd size has been increasing over the years, so has individual cow yield. In the 1940s it was only 3000 litres per cow per year whereas the average yield is currently 5300 litres... with some cows now producing over 10,000 litres per year' (Carter and Stansfield, 1994:41).

Developments such as these can have beneficial financial implications for the producer, although it must be recognised that at the same time there can be detrimental welfare implications for the stock. These may not be a direct consequence of that animal becoming more productive, however, but may well result from the stock keeper not appreciating the consequential nutritional, housing and management needs of a highly productive animal (Webster 1994:141). 'The large, genetically superior dairy cow consuming a ration based largely on grass silage may suffer both from hunger and pain, or at least chronic discomfort, in a cubicle house partly because the quality of feed has become inadequate to meet her nutrient requirements for lactation and she has lost condition, partly because the wet silage has contributed to poor hygiene and predisposed to foot lameness, and partly because genetic selection has created a cow too big for the cubicles' (Webster, 1994:143).

The origins of the problem are not to be found the science of livestock breeding but in the keeper's lack of understanding of the animal, its nature and needs, in his irresponsible management of his stock, and in his poor standards of stockmanship. In other words, the animal's welfare problems are directly related to an ignoring of the principle of respect. I suggest, therefore, that research into breed improvement should pay due regard to this principle, recognising that it provides the ethical basis for continued and responsible breed development. As FAWC (1999:9) have commented, 'there is an urgent requirement for research aimed at assessing the welfare of different breeds of stock, and their suitability for the environment in which they are kept. Such research should incorporate a wide range of welfare indicators, and the results must be used to influence the direction of future breeding programmes to the benefit of the animals' welfare.' Not only do I endorse FAWC's recommendation regarding the need for such research but would also argue that the research should incorporate the respect principle as previously defined as a means of assessing the ethical as well as the welfare standards of intensive production systems.

If this were to happen, and the respect principle was accepted as the ethical base for livestock production systems in the UK, and incorporated into those further areas for agricultural research previously prioritised by FAWC (1993:6-10), I believe this would provide essential criteria and support for judgements to be made regarding the acceptability of production methods. As the principle refers to respect for an animal's essence, for everything that makes it what it is, so those research areas identified by FAWC as 'Perception and Cognition' should provide new opportunities for deepening our understanding of every aspect of an animal's nature and character, its essence. Pursued through a variety of scientific disciplines, from animal psychology to physiology and ethology (Webster, 1994:10), such research should enable us to learn how an animal experiences its environment, what and how it feels under a variety of circumstances and how those feelings are likely to be manifest. In this way we approach the attaining of Webster's dream (1994:10), that he might be able to see the world through an animal's eyes.

The implications are correspondingly great. As FAWC (1993:9) have commented, the achieving of this level of awareness should serve to 'increase our own humility as to the limits of our ability to understand the minds of farm animals and increase our own motivation to improve that understanding.' These sentiments are very much in accord with the whole tenor of this thesis, especially when the expressed desire to improve our understanding of an animal's mind extends to all aspects of their nature and character. In this way we will not only discover more about the very essence of that animal but will be more able to assess whether a husbandry system or method offends the principle of respect by ignoring or debasing what that animal essentially is.

An example of the relevance of the above for a commercial livestock environment is found in research carried out by Carson and Wood-Gush (1984) into the behaviour of calves at markets. Observation of the immature animals' varied reactions to the presence or absence of stimuli in the market environment allied to similar observations of calves studied in control conditions, enabled the researchers to interpret the meaning of the different behaviour patterns. For example, 'at the real markets behavioural signs of distress were apparent and initially these included loud vocalizations, walking or running about while looking around with eyes staring, ears pulled back and tails clamped down' (Carson and Wood-Gush, 1984:392).

From such observations and from comparisons between the two groups of calves it was possible to arrive at conclusions which led in turn to the making of a number of recommendations regarding the way in which calves should be handled at market. Understanding of and subsequent respect for the feelings of the calves had resulted in improved standards of welfare in the market environment. In terms of the principle of respect, understanding has been expressed through improved management and care of the stock, so satisfying the principle's established criteria.

It is also worth noting that this research project again illustrates the correlation that is often found between welfare considerations and commercial benefit. The factors that were causing the calves to feel distressed and frightened in the livestock market were identified as being the cause of their higher mortality rate compared with that of calves reared at their place of birth. Whilst the risk was identified as being 1.6 times greater (Carson and Wood-Gush, 1984:389), the researchers' recommendations for improved handling techniques and for a greater sensitivity of the animals' needs and feelings should remove the fatal stressors and subsequently lead to a reduction in that mortality rate and to an increase of the rearers' profitability.

The practical application of the conclusions and recommendations arising out of research such as that outlined above can clearly have benefits for animal and producer. I suggest that this remains the case even when there is no observable financial gain. As was observed at the start of this chapter, there can be a great deal of frustration for the dedicated stockperson trying to apply the principle of respect to their enterprise in having to employ practices and methods that run contrary to that ethic. Conversely, when the necessity of employing such methods is removed, then the frustration can be replaced with a renewed sense of professional and personal satisfaction. Any scientific research that facilitates that process is to be welcomed and encouraged.

7.5   Training

As a rule people adopt ethical principles through assimilation, information or compulsion and frequently a combination of all three. The first occurs as a number of factors such as home background and peer environment direct the early years of childhood or adolescence and cause an individual to behave and act in a manner that is acceptable in their social grouping. Imperceptibly and often unknowingly the child absorbs the underlying ethical principles of their society. In this way, the guidance and influence of companions who are trusted and looked up to, be they parents, teachers, or friends, alongside the overall early experiences of a person's ethnic culture, tend to establish the beliefs and principles of the early years. At the same time this process often produces the framework around which that person's adult ethical system eventually takes shape and their convictions are formed.

This next stage in a person's moral development occurs either through the individual unquestioningly accepting that which has previously been assimilated, or as a result of their exposing these same ideas and beliefs to critical examination. Whichever is the case, the adult now holds to and lives by a set of ethical principles which they believe to be right. The latter process of examining and analysing one's early beliefs involves the imbibing of alternative ideas, and the analysing of other belief systems. It needs intellectual and philosophical or spiritual stimulation, often all three and this I suggest requires a process of education which ultimately leads to the formation and adherence to personal moral convictions.

The third formative factor, that of compulsion, requires the imposition of regulations and rules, with appropriate punishment being administered if those rules are broken. This process need not be legalistic, although the fear of breaking a law and facing due retribution can cause a person to live according to certain ethical principles which they need not necessarily believe in. I may not be convinced, for example, that theft is wrong, but a dread of being caught and punished causes me to resist the temptation to steal. Similarly, the fear of being ostracised by my peer group or of facing the condemnation of the wider community may persuade me to live by certain principles that again I do not hold by conviction. So, for example, out of a desire to keep my job, I may uphold the principle of gender or racial equality in the work place without necessarily believing in either. In this way legal or social compulsion may result in my living by certain moral principles. Whether I hold those principles by conviction is practically irrelevant.

Clearly, all three of these factors regarding the promotion and adoption of ethical principles have a bearing on the promotion and application of the principle of respect in the world of livestock husbandry. Indeed, within the farming community assimilation has often been seen as the way in which the principles of stock management and stockmanship have been passed on through the generations (English *et al.*, 1992:98). This is often reflected in the comments which suggest that the aptitude and skill required by stock keepers 'are in the blood' or 'passed on in the mother's milk', or that 'stockmen are born not made.' Whilst assimilation can undoubtedly be a strong factor in producing well-motivated and responsible stock keepers who live and work with their animals according to the principle of respect, similar results can be obtained through training and educating individuals in these areas, providing they have the personality and temperament to handle animals.

I believe that this recognition of the role that training can have in promoting ethical principles in farming has important implications for those responsible for establishing the strategy and priorities for agricultural education. Indeed, I would argue that any training in the technical and practical aspects of animal husbandry should be balanced by a similar emphasis on the ethical principles that direct the way in which we care for livestock. As English *et al.* (1992:97) have suggested, such courses should include 'the specific goal of influencing personal attitudes to the animals so as to improve their care.' Brooman and Legge (1997:434) have taken this suggestion further and have expressed it in the terms already adopted in this thesis, as they argue that 'education to promote the necessity to treat animals with respect has importance in the initial schooling of all individuals and is particularly relevant to the initial training of all those who work with animals.'

I believe that, when placed within the practical world of animal husbandry, such training enables the students to appreciate how the welfare of their stock is influenced by their own ethical stance concerning that stock. Furthermore, as I have previously argued, I believe that the principle of respect for the essence of one's animals, expressed through responsible stock management and high standards of stockmanship, provides the ethical base from which such training can be developed.

I suggest that the introduction of this element into the training of those intending to embark on a career in livestock production could have major implications for the future well-being, not only of the stock and the stock keeper, but also of the industry itself. I believe not only that it would equip those stock keepers at a personal level with the ability to examine and understand the ethical issues relating to their work, but also by their applying the respect principle's criteria to that work they would then be able to establish ethically acceptable practices and methods across the complete spectrum of livestock production.

A factor likely to influence this, however, is that over recent years the number of full-time livestock producers in Britain has been decreasing. In part this has been due to the economic crisis in the industry, although it is also a result of the greater vocational freedom and choice enjoyed by society. The consequence of both is that the next generation from farming families are less inclined to follow the traditional pattern of following their parents into a career on the land. The impact that this is likely to have on the agricultural community was indicated in the Social Audit carried out by the NFU in the summer of 1999. The researchers discovered that '66% of farmers' children do not intend to take over the farm after they retire, raising the prospect of farms being deserted by disillusioned future generations' (NFU Sept. 1999).

These social changes must be taken into account when discussing training for livestock keepers, for this decline in the number of full-time producers has been more than matched by a steady increase in the number of part-time and hobby farmers. These newcomers to the industry will not have had the benefit of the traditional process of agricultural training whereby farming culture and skills were assimilated by the next generation of farmers through the atmosphere and ethos of home. As a result, those responsible for agricultural training must begin to take the needs and circumstances of this new clientele into account when formulating their strategies for the future of agricultural education.

Moreover, the likelihood is that the number of part time and hobby farmers is going to continue to increase in the foreseeable future. Indeed, from the annual statistic produced by MAFF for registered holdings in the UK, the NFU have predicted (Table 19) that whilst the number of full-time farmers is likely to fall by 13% in the period 1995 to 2005, the number of part-time farmers is predicted to increase by 18% (NFU, 1998:25).

Table 19   **Current and predicted agricultural labour force in England and Wales**
(Adapted from NFU, 1998:107)

Year	Full-time farmers, partners and directors ('000)	Part-time farmers, partners and directors ('000)	Total ('000)
1985	154	57	211
1990	142	59	201
2000	123	74	197
2005	115	78	193

I suggest that, whilst a sizable proportion of the growing number of part-time farmers may have previously managed their holdings on a full-time basis and others will have always run their farms in this way, the remainder will consist of individual who are total newcomers to agriculture. Attracted by the vision of a rural life-style and the prospect of becoming at least partially self-sufficient in food, it is likely that they will have had little in the way of training in animal husbandry either formally through an agricultural college, or informally through family background and assimilation of knowledge and skills. On the same basis I would argue that they are unlikely to have had training in the ethical issues and concerns relating to the world of livestock production.

Whilst these new-entry hobby farmers are an easy target for the humour of well-established rural dwellers, including members of the agricultural community (Brewis, 1992), I would argue that they can be a major asset to the prosperity of the countryside and the future well-being of rural communities in general and farming in particular. From my personal life-time's experience of living and working in various rural parts of England I have found that these newcomers to farming can often bring an enthusiasm and commitment to livestock matters that those already in the industry have lost either through familiarity or as a result of the current crisis. Furthermore, I suggest that, as they have chosen to live in a rural environment and appreciate its many benefits, they consequently want to put down roots and establish a long-term commitment to the community in which they are based.

Personal experience has shown that new part-time farmers are frequently from well-educated and prosperous backgrounds and are able to afford and keen to consume books and articles on matters such as livestock management as a means of preparing themselves to care for their animals. I have also found that they are keen to seek advice on matters such as animal health and nutrition and are enthusiastic to learn how to cope with difficulties such as awkward lambings or dealing with foot rot or fly strike. It is, therefore, of some importance that they should have access to sympathetic advice and support from their farming neighbours as well as a supply of reading material that will deal with the ethical, as well as the practical, principles regarding livestock production and management.

The organisations to which many of these hobby farmers will belong, such as NFU Countryside and the Rare Breeds Survival Trust, clearly have a major role in providing this support and training. Indeed, the 'Country Guides' produced by NFU Countryside as inserts in the 'Countryside Folder' which is given to their members, are an excellent example of the material that is required (NFU Countryside, 2000). The Guides are practical and simple, providing an insight for any newcomer into the needs and character of the animals they are thinking of farming. Advice is given on legal matters regarding the keeping of stock, and the welfare and moral obligations that ownership of animals places on each stock keeper is also highlighted. The practical application of animal welfare is firmly based on FAWC's Five Freedoms, and my only criticism would be the omission of any mention of the ethical principles that undergird these obligations. I suggest that guides such as these from the NFU should also introduce their readers to the respect principle and its criteria as the ethical base for the livestock enterprise.

The 'Countryside Guides' also encourage newcomers to enrol in short courses provided by Agricultural Colleges and LANTRA ('Landbased Training') and to join those organisations which can also offer training and advice, such as the breed societies, or organisations such as The Rare Breed Survival Trust. Advice is also given regarding publications worth subscribing to, such as 'The Smallholder' or 'Farmers Weekly'. I suggest that, through the inclusion of well-argued articles concerning the ethical base for livestock production, newcomers could be encouraged to accept and apply the principles to livestock activities on their holdings. I would also argue that the respect principle, as outlined in previous chapters and expressed through its criteria of responsible stock management and the exercising of high quality stockmanship, provides the ideal ethical base for livestock enterprises large or small, extensive or intensive and that it should therefore be promoted in this way. It is to be hoped that as a result hobby as well as commercial farmers will come to a greater understanding of the nature and needs of their stock, and therefore be able to respect fully the essence of those animals.

As organisations, training courses and publications are available to all those engaged in livestock production, so too they are all governed by the same legislation regarding the welfare of their stock. Just as there can be a moral aspect to these other influences, so I would argue that the legislation regarding the keeping of animals should similarly be built on a firm ethical base.

This has never been easy and in the multi-cultured legislative world of the European Union (EU) it has proved to be particularly difficult. Brooman and Legge (1997:190) have commented that 'as economics, not morality, is the cornerstone of the EU, the issue of animal welfare has always settled a little uncomfortably within the original economic objectives of the Treaty (of Rome).' This tension is later illustrated in their statement that 'animals involved in the production of meat and other animal products are the subject of a commercial trade-off between the demand for cheap produce and a growing recognition that animals have a need for more protection by the law' (Brooman and Legge, 1997:207).

It would appear from comments such as these, that contrary to current practice, there is a need for greater ethical input into all debate concerning farm animal welfare legislation, nationally and internationally. If this were to happen, then I believe that it could have a major impact on the current legislative approach to farm animal welfare, in that it could move the debate forward from issues concerning the avoidance of cruelty to the active promotion of measures that improve welfare (Everton, 1989:102). Indeed, Everton (1989:103) suggested that central to any new legislation arising out of this more ethically based debate 'would be a widely phrased duty to positively promote the well-being of farm livestock.' As FAWC's Five Freedoms and their various welfare codes become more widely acknowledged as a practical welfare basis for animal legislation, one can see this move already beginning to take place. I suggest, however, that alongside the Five Freedoms, the principle of respect provides an ideal ethical reference point for the encouragement of such debate. Focused on an awareness and appreciation of an animal's essence and emphasising the importance of responsible management and skilled stockmanship it actively promotes and enhances livestock welfare. Indeed, in promoting the attainment of the highest possible standards of animal systems management and stockmanship skills, the principle fulfils one of Everton's main concerns as she argues that 'one fundamental principle of humanity (is) that man as keeper of the animal is obliged to protect it to the BEST of his ability' (Everton, 1989:107).

The use of the principle in this way as a reference point for legislative debate will not only endorse the potential benefits its application has for livestock but will also promote its wider adoption, in that those livestock producers who do not hold to the principle by conviction would have to adhere to it by compulsion. As Brooman and Legge (1997:173) appreciated 'a powerful aversion to animal cruelty - that is wilfully and uselessly harming an animal, or harming an animal for frivolous reasons - is ingrained in virtually all agriculturalists, especially those who have come from an extensive background.' There has always been a small minority, however, for whom ethical concerns have been of little significance and it is these who will be encouraged in this way by law to adopt and apply the principle of respect on their holdings even when they fail to do so by conviction.

CHAPTER 8

Conclusions

The aim of this study was to investigate the ethical issues in intensive livestock production so that, if possible, a framework could be developed to encapsulate those principles of human behaviour and so provide the ethical criteria for assessing intensive production methods.

As was recognised in Chapter 2, many of those opposed to current methods of livestock production have focused their criticisms on the process of intensification itself, and especially its impact on animal welfare (Harrison, 1964). Whilst appreciating and sharing their concern for the welfare of farm animals, I argued that their criticisms have often been based on a broad and often unwarranted acceptance of an animal rights and vegetarian philosophy (Chapter 4).

I proposed the principle of **respect for an animal's essence** as a practical ethical basis for the human/animal relationship and as an alternative to the concept of animal rights (Chapter 5). Respect is expressed in valuing and accepting responsibility towards that which is to be respected (Rolston:1988). I suggested that in terms of livestock production this **is equated with the concepts of stockmanship and stewardship**.

As a result, I argued that the principle of respect so understood, provides an ethical basis for intensive livestock production systems and that the principles of stockmanship and responsible stock management provide the criteria by which the ethical status of such systems can be assessed (Chapter 5). On this basis I concluded that whilst some current practices are unacceptable (e.g. beak clipping) others are not (e.g. robotic milking). Indeed, it was found that when the principle's criteria of stockmanship and stewardship were applied to certain current intensive systems they were found to be more beneficial to animal welfare than some of the rearing systems of the past (e.g. the rearing of individual pigs in small sties).

279

8.1 Debating the principle of respect

Of all the factors that influence human behaviour and belief, debate is arguably one of the most effective. Combining the power of reasoned argument with the persuasive passion of vocal intonation or the carefully constructed use of the written word, debate can stimulate analysis and reflection and clarify understanding. In this respect, the area of legislative debate mentioned in Section 7.5 serves as a vivid example of this whole process, combining as it does the impassioned plea of the courtroom with the carefully phrased textbook on law.

Unfortunately, the debate on livestock production systems and the use of farm animals for human purposes can easily generate a great deal of passion without producing much clarity or understanding. This may well be due to the confrontational and prejudiced attitude that the two sides often seem to adopt towards each other (Leahy, 1991:254 and 258). The implications for any livestock farmer who reacts to his critics in this way can be serious and far-reaching, as his dismissal of their genuine concerns over production methods may be harmful to his own interests. In refusing to listen to consumer concerns he runs the risk of losing their custom and with it his means of making a living. Similarly, in failing to make a well-argued response to criticisms from the vegetarian and animal rights lobbies, he courts the danger of those same groups persuading others to reject the consumption of animal products. This clearly has the same outcome as the earlier scenario.

Consequently, I believe that the future of the livestock industry depends on its representatives encouraging people of good will from both sides to meet and talk together. Through this process of listening to and learning from each other, points of common agreement should emerge which might in turn lead to the resolution of some of the concerns regarding production systems and methods. A serious forum should be established to facilitate understanding and respect between the farming community on the one hand and the advocates of animal rights and liberation, for example, on the other. As has already been suggested in Section 4.5 there are organisations, such as UFAW or CIWF, which are ideally placed to carry this forward because of the wide range of interests represented in their membership and who need to be encouraged to do so.

Whilst each side may be motivated by very different views concerning the moral standing of animals, I argue that the two groups share a common ground, in that each is motivated by, and is ready to express a concern for those creatures and to care deeply about them. Building on this foundation the two sides working together should be able to establish what the needs are of the different species of farm animals and how those needs should be met. This is an important next step in the application of the respect principle, and might form the basis of future research, sponsored by the interested parties. From this work, it will hopefully be possible to reach agreement on what, if any, improvements are required to the methods and systems currently being used by UK livestock producers.

Furthermore, I contend that the principle of respect, as presented and discussed in this book, could be the ideal starting point from which discussion between the two sides could develop, as it provides a point of common interest between the different parties. In fact, in agreeing that respect should be the basis of our relationship with animals, the two opposing sides might well be encouraged to adopt the principle of respect as the basis for their own relationship with each other. A forum could be developed to include representation from other groups concerned with the human/animal relationship. Concerns about the keeping of animals in zoos, their use in medical research, or their possession as pets, for example, might well be addressed in this way. As a result, the principle of respect could be developed further and applied to other issues, leading to its wider adoption not only in the livestock industry but right across society.

8.2 Developing respect

If the above recommendation is to have any positive outcome for the farming community it requires that the views of livestock producers, and their guiding principles be well founded and equally well argued. Indeed, the suggestion was made in Section 2.6 that livestock producers need to be trained in the art of apologetics. To that end I suggest that all agricultural education should include the study of those ethical issues related to the systems of livestock production.

The study of ethical issues would not only equip livestock producers to engage in the debate on these matters, but would also encourage them to examine their own relationship with and attitude towards their stock. The promotion and application of the principle of respect provides the ideal foundation on which this training can be built.

In addition, it was suggested in Section 7.5 that such training should also have a strong practical emphasis. It was argued that the stockmanship element of the respect principle requires the stock keeper to have a deep understanding of the animal's essence and possess the ability to empathise with the animal's needs. As a result, a trainee not only needs to learn about the insights of ethology and biochemistry into the world as the animal sees it, but also needs to learn how to observe and assess the needs of the animals. In previous generations this was referred to as 'walking stick farming', whereby the stockman would lean on his stick and just watch over his stock - a practice that some new innovations in livestock husbandry, such as automated milking, may well reintroduce and encourage.

There are two particular areas regarding the provision of training, however, where there are problems to be resolved. One relates to the conflict which many stock keepers are currently experiencing. Financial pressures have caused the industry to shed labour at an unprecedented rate (MAFF, 2000:2.10). Those who are left are forced to carry an increased workload. At the same time, the drive for greater efficiency has brought about an overall increase in the size of livestock units (MAFF, 2000:2.8/9). Time and finance for further training, no matter how well motivated and enthusiastic the stock keeper may be, has become negligible. This is particularly so for those who are self-employed and who work on their own. I would argue, therefore, that if producers must keep abreast of developments in the areas of ethics, ethology and biochemistry as well as those of technology, then the means must be found for them to continue learning.

One way in which this could be achieved would be through the provision of government funding. This would not only meet any training costs, but also cover any expenses incurred in employing relief labour to deal with such essential tasks as feeding or milking, whilst the stock keeper is off the farm. Another would be the inclusion of discussion on these issues in the programme of local NFU branch meetings or farm discussion groups, and in the pages of the farming press. This rarely happens at present.

The other concern with regard to training provision relates to the needs of new entry part-time farmers. It was stated in Section 7.5 that this is a growing sector of the industry that contains many entrants with little farming knowledge or experience. A limited response to their training needs is currently being made by organisations such as NFU Countryside through the provision of an array of informative material. In a similar vein some Agricultural Colleges offer relevant evening or weekend courses. The temptation is for both to focus on training in technical and practical skills without considering the ethical debate or empathy as an element of stockmanship.

I believe that the promotion of the respect principle in this sector is essential, especially as there may be the tendency for some new part-time entrants or hobby farmers to either treat their stock as pets or to have an idealised view of the stock rearing systems of the past. Both of these approaches to livestock husbandry could contravene the respect principle and be detrimental to the welfare of the animals.

8.3 Future changes

If general acceptance of the respect principle as the ethical basis of livestock production systems were to be achieved, then I would argue that certain actions must then follow. For example, a number of practices or systems currently being used would become unacceptable as that they do not comply with the criteria of quality stockmanship and responsible stock management that the principle requires. Indeed, one example of what I believe to be an unacceptable practice was discussed in Chapter 6, where it was argued that the practice of beak clipping is no longer an acceptable way of dealing with the problems of feather pecking and cannibalism, which are problematic in a number of poultry rearing systems. The rejection of the practice was based on its failure to respect the essence of the birds, in that it frustrated their natural instinct to explore their environment with their beaks and that it caused them long term suffering.

Other practical implications of the principle being accepted as the ethical basis of the livestock industry also need considering. Sainsbury (1986), for example, recognised that the design of stock housing and handling equipment can have a great impact on animal welfare. If, as has been suggested in Chapter 6, designers were to take account of the principle of respect and its emphasis on the essence of the animal then the starting point for the design of both housing and handling equipment would be the needs of the animal concerned rather than human convenience, or economic considerations. This would have clear benefits for the welfare of the stock concerned.

I suggest also that the relevance of the respect principle goes beyond the world of the livestock producer and that it has implications for retailers and consumers. For example, the potential value of Farm Assurance schemes was recognised in terms of farm animal welfare and product identity, but the confusion arising out of the plethora of different ethical and welfare criteria on which the schemes are based was also recognised. Some of this confusion could be resolved by the widespread adoption of the respect principle by supermarkets, animal welfare organisations and farmers as the ethical basis for the various assurance schemes which deal with the sale of animal products. With its emphasis on high standards of stockmanship and the responsible management of stock, as well as its focus on respect for the essence of the animals and all that that entails in terms of appreciating and responding to their needs, I believe that the principle could find acceptance amongst all the parties concerned.

8.4 Areas for further research

In establishing the principle of respect as the ethical basis for intensive livestock production systems, a number of areas relevant to the study were identified for further exploration. This study for example, focused solely on commercial livestock production in the UK and in the light of the respect principle assessed the ethical acceptability of a range of practices and systems commonly used by UK farmers.

British agriculture competes, however, on a world market and is also an integral part of the European Communities agricultural scene. As a result, the wider trade implications arising out of the application of the respect principle by UK producers also need to be examined.

The impact that such an acceptance might have on the competitiveness of British animal products for example, is a particular issue that needs considering, as is the possibility of other countries adopting the principle as the ethical basis of their own livestock industries. Many nations continue to use animal rearing practices that this country has now made illegal, such as sow stalls and tethers or the docking of cows' tails, whilst others have developed new systems, such as the vast hog farms and beeflots in parts of the USA, that I believe would be unacceptable under the terms of the respect principle. I hope that wider consideration of the principle's usefulness and practicality as well as the sound ethical arguments in its favour will encourage other nations to apply it to their own livestock systems.

Just as the response of other nations to the promotion of the principle remains unknown, so there are aspects of the study with regard to its adoption and application in the UK that warrant further exploration. In this study, for example, all livestock keepers have been considered as one body. In reality they are made up of employers, employees and the self-employed, some working part time, some full time, but all working with stock and having some degree of responsibility for the care of their animals. In the course of my studies I have been unable to discover any research into possible differences of motivation, standards of care or sense of responsibility that these different categories might exhibit in their relationships with their animals. These are factors which could have a bearing on how the individuals concerned accept and apply the respect principle, and they warrant further exploration. Some sectors of the industry, for example, such as egg and broiler production, depend on a large employed work force whose relationship with their birds and whose standards of stockmanship may well be very different to those of the self-employed dairy producer who owns and knows each one of his cows. No study appears to have been done on these differences and their implications for the world of stock rearing. This area warrants further research.

Similarly, as there appears to have been no research into the ethical basis of Farm Assurance Schemes or recent legislation regarding farm animal welfare. Adoption of the respect principle provides the motivation for this assessment to take place.

In focusing on commercial animal production in the UK, this study has had to omit concerns related to the other sectors of the British livestock industry. Issues such as the sale and transportation of stock and especially the slaughter of animals have been particularly contentious over recent years. I believe that the respect principle offers an ethical basis for assessing the treatment of animals in each of these areas, and would suggest, therefore, that the application of the principle in these other sectors of the industry should be considered.

8.5 The case for respect

Having spent a large part of my life working with livestock and having pursued this course of study I now believe firmly that the principle of respect, as it has been presented here, provides the industry with a practical ethical base for commercial production systems. In expressing that respect for an animal's essence through responsible stock management and committed stockmanship I also believe that it upholds all that is best regarding the traditions of animal care to which many British farmers have long adhered and have justifiably been proud of. I also believe that this conviction is merited by the arguments proposed above and that the principle can be applied to all animal production systems, be they intensive or extensive.

Furthermore, having spent a large part of my life living and working with people who breed and rear livestock, I firmly believe that many of them already farm in this way. Indeed, I submit that the reason why the Lincolnshire farmers referred to in Chapter 1 expressed such anger and outrage at the poor condition of a sheep being sold at Louth market was because of their implicit belief in the principle of respect. It is to these men that I dedicate this book in the hope that it explains their concern for one elderly sheep and reflects the attitude of responsible stockmen towards all stock.

Bibliography

Aaker, J. (Ed.) (1994) *Livestock for a Small Earth* Heifer Project International, Washington, DC

Abbott T., Hunter E., Guise H., Penny R., (1996) Survey of farrowing management of outdoor pig production systems *Proceedings of the 14th. International Pig Veterinary Society Congress* Bolgna, Italy 7th - 10th July 1996:627

Adam, D. (1985) *The Edge of Glory* SPCK, London

ADAS, (1997) *Agricultural Strategy 1997* ADAS, Kiddlington, Oxon

Addy J., (1972) *The Agrarian Revolution* Longmans, London

Anderson B., (1957) *The Living World of the Old Testament* Longmans, London

Anon (Summer/Autumn 1994) *AgScene* Compassion in World Farming, Petersfield

Anon (Spring 1995) *Agscene* Compassion in World Farming, Petersfield

Anon (Autumn 1996) *AgScene* Compassion in World Farming, Petersfield

Anon (Summer 1997) *Agscene* Compassion in World Farming, Petersfield

Anon (Spring 1999) *AgScene* Compassion in World Farming, Petersfield

Anon (April 1997) Farm Assurance *British Farmer and Grower* NFU, London

Anon (July 1997) 'New Live Transport Rules *British Farmer and Grower* NFU, London

Anon (2000) Bishop warns of agricultural meltdown *Church Times* 3rd. March 2000

Anon (Oct. 1997) *Foresight for Food and Drink - meat* DTI, London

Anon (May 1998) *Foresight for Agriculture, Horticulture and Forestry - Livestock Production* DTI, London

Anon (1996) Growing Pains *The Economist* 20th April 1996

Anon (1997) Robotics under review *Farmers Weekly* 1st. Aug. 1997

Anon (1998) Free range broilers could be a good bet *Farmers Weekly* 5th. June 1998

Anon (1998) Dairy first for RSPCA *Farmers Weekly* 8th. May 1998

Anon (1998) Readers Letters *Farmers Weekly* 11th. Sept. 1998

Anon (1998) *Farmers Weekly* 18th. Sept. 1998

Anon (1998) UK wants world welfare standards in WTO talks *Farmers Weekly* 2nd. Oct. 1998

Anon (1998) *Farmers Weekly* 10th. Nov. 1998

Anon (1998) Antibiotic restrictions are urged *Farmers Weekly* 11th. Dec. 1998

Anon (1999) Lameness checklist may cut incidence *Farmers Weekly* 26th. March 1999

Anon (1999) Locomotion scoring tells all *Farmers Weekly* 26th. March 1999

Anon (1999) National disposal plan needed to end misery of unwanted calves. *Farmers Weekly* 20th. Aug. 1999

Anon (1999) Feeding tactics curb aggression in loose housing *Farmers Weekly* 29th. Oct. 1999

Anon (1999) *Farmers Weekly* 19th. Nov. 1999

Anon (1999) Dutch pig men to phase out AGPs *Farmers Weekly* 3rd. Dec. 1999

Anon (1999) Fewer units – more output *Farmers Weekly* 31st. Dec. 1999

Anon (1999) Loose housing no loss *Farmers Weekly* 31st. Dec. 1999

Anon (January 1994) Chops off the Old Block loses Appeal *Financial Times*

Anon (July 1998) How Cheap Imports are Devastating Poultry Jobs *Landworker* T & G Publications, London

Anon (May 1999) Welfare concerns link new EU BST reports *NFU Business* NFU, London

Anon (June 1999) Council Agreement on Layer Welfare *NFU Briefing* NFU, London

Anon (July 1999) *NFU Business* NFU, London

Anon (August 1999) Hen Directive will cost UK more than £200m *NFU Business* NFU, London

Anon (Jan. 2000) UK Poultry sector stares into abyss *NFU Business* NFU, London

Anon (mid-March 2000) Contract with Society eagerly awaited *NFU Business* NFU, London

Anon (March 2000) O'Brien Bill will stop food label origin "hoodwinking." *NFU Business* NFU, London

Anon (March 2000) Supermarket probe plea *NFU Business* NFU, London

Anon (March 2000) Tesco makes new pledge to farming *NFU Business* NFU, London

Anon (April 2000) British Farm Standard food label launched *NFU Business* NFU, London

Anon (Winter, 1998) Farmers Work to Win Trust *NFU Magazine* NFU, London

Anon (April 1998) Ministry vets to run checks on tail docking routines *Pig Industry*

Anon (Sept. 1998) The Survey *Pig World*

Anon (Dec. 1999) New Report nails their animal welfare claims *Pig World*

Anon (Spring 1999) FABBL/ABM initiative begins on 1 April *The Standard* Farm Assured British Beef and Lamb, Milton Keynes

Anon (1999) Is Frankenstein Food bad for us? *The Week* 20th. Feb 1999

Archbishop of Canterbury (1981) *Statement on Animal Welfare Matters* Church House Publishing, London

288

Archbishop's Commission on Rural Areas (1990) *Faith in the Countryside* Churchman's Publishing, Worthing, Sussex

Arey D., and Bruce J., (1993) A note on the behaviour and performance of growing pigs provided with straw in a novel housing system *Animal Production* **56**:269-272

BBSRC (1996) *The New Biotechnologies – opportunities and challenges* Biotechnologies and Biological Research Council, Swindon

Bailey L.H., (1988) *The Holy Earth* The National United Methodist Rural Fellowship, Colombus, Ohio

Bennett R., (1995) The value of farm animal welfare *The Journal of Agricultural Economics* **46** (1):46-60

Bennett R. and Larson D., (1996) Contingent valuation of the perceived benefits of farm animal welfare legislation: an exploratory survey *The Journal of Agricultural Economics* **47** (2):224-235

Bennett R., (1996) Willingness-to-pay measures of public support for farm animal welfare legislation The *Veterinary Record* **139**:320-321

Bennett T., (March 2000) EU's vitamins-in-feed plan is "madness" *NFU Business* NFU, London, UK

Beresford T., (1975) *We Plough the Fields* Pelican Books, Harmondsworth, Middx.

Blackman, Humphreys and Todd, (1989) *Animal Welfare and the Law* Cambridge University Press, Cambridge

Bonham-Carter V., (1971) *The Survival of the English Countryside* Hodder and Stoughton, London

Bonhoeffer D., (1970) *Ethics* Fontana, London

Bourns C (18th. June 1999) *Council Agreement on the Welfare of Laying Hens* NFU, London

Brambell, F.W. Rogers, (Chairman) (1965) *Report of the Technical Committee to Enquire into the Welfare of Animals kept under Intensive Livestock Husbandry Systems* HMSO, London

Brewis H., (1992) *Country Dance* Farming Press, Ipswich

Briggs, A (1983) *A Social History of England* Weidenfeld and Nicolson, London

Britton, et al (1983) *The Changing Farm* Arthur Rank Centre, Kenilworth

Broom D., (1989) *Ethical Dilemmas in Animal Usage* in *The Status of Animals* ed.d by Patterson D., and Palmer M. CAB International, Wallingford

Brooman S. and Legge D., (1997) *The Law Relating to Animals* Cavendish Publishing, London

Brownlow M., Carruthers S., Dorward P., (1995) Financial Aspects of Finishing Pigs on Range *Farm Management* **9**(3):125-132

Bruce A and D., (1998) *Engineering Genesis* Earthscan Publications Ltd., London

289

Brunner E., (1947) *Man in Revolt* Lutterworth Press, London
Budiansky S (1994) *The Covenant of the Wild* Weidenfeld and Nicolson, London
Burrough, D. (Jan. 1999) Direct Action Works *Pig World*
Buss J., (27th. November 1998) Be prepared as antibiotic ban looks imminent *Farmers Weekly*
Carnell P., (1983) *Alternatives to Factory Farming* Earth Resources Research Ltd., London
Carpenter E., (1980) *Animals and Ethics* Watkins Publishing, Dulverton, UK
Carruthers J., (1990) *Farm Animal Welfare : It Pays to be Humane* Farm Animal Care Trust, London
Carruthers S., (1991) *It Pays to be Humane* Centre for Agricultural Strategy, University of Reading
Carruthers P., (1992) *The Animals Issue* Cambridge University Press, Cambridge
Carson K., and Wood-Gush D., (1984) The Behaviour of Calves at Market *Animal Production* **39**:389-397
Carter E, and Stansfield M, (1994) *British Farming: Changing Policies and Production Systems* Farming Press, Ipswich
Chamberlain D., (1991) Changing the Image Farming and Farmers *Farm Management* 7(10):477-483
Church of Scotland (1986) *While the Earth Endures* Church of Scotland Department of Ministry and Mission, Edinburgh
Clarke S., (1997) *Animals and their Moral Standing* Routledge, London
Clough, C., and Kew, B. (1993) *The Animal Welfare Handbook* Fourth Estate, London
Clutton-Brock J., (1994) The Unnatural World – Behavioural Aspects of Humans and Animals in the Process of Domestication *Animals and Human Society* (ed. Manning and Serpel J.) Routledge, London
Colman D., (1994) Ethics and Externalities: Agricultural Stewardship and Other Behaviour *Journal of Agricultural Economics* **45**(3):299-311
Dawkins, M. (1988) *Animal Suffering - The Science of Animal Welfare* Chapman and Hall, London
Dawkins, M. (1993) *Through Our Eyes Only* Freemans London
De Grazia, D. (1996) *Taking Animals Seriously* Cambridge University Press, Cambridge
Deloitte & Touche Agriculture (1999) *Farm Results 1998/99* Deloitte and Touche, London
Donaldson J and F., (1972) *Farming in Britain Today* Pelican Books, London
Donnellan C., (ed.) (1994) *The Vegetarian Choice* Independence Educational Publishers, Cambridge

Dower N., (1996) Global Hunger: Moral dilemmas *Food Ethics* (ed. Mepham B.) Routledge, London

Drew B., (1996) New Technology – Livestock *Farm Management* 9(5):251-259

Eddison, J. (1995) Animal Welfare *Primrose McConnell's The Agricultural Notebook* (ed. Soffe, R.), Blackwell Science, Oxford

English P., Burgess G., Segundo R., Dunne J., (1992) *Stockmanship. Improving the Care of the Pig and Other Livestock* Farming Press, Ipswich

Evans P., (1996) *Farming for Ever* Sapey Press, Whitbourne, Worcester

Everton A., (1989) The Legal Protection of Livestock *Animal Welfare and the Law* (ed. Blackman D., Humphreys P., and Todd P.) Cambridge University Press, Cambridge

Ewart Evans G., (1956) *Ask the Fellows Who Cut the Hay* Faber Ltd., London

Ewbank R., (1994) Animal Welfare *Management and Welfare of Farm Animals'* UFAW, Potters Bar, Herts

Farm and Food Society, The (1972) *An Enquiry into the effects of Modern Livestock Production on the Total Environment* The Farm and Food Society, London

FAWC (1981) *Advice to the Agricultural Ministers of Great Britain on the Need to Control Certain Mutilations on Farm Animals* MAFF, London

FAWC (1991) *Report on the Welfare of Laying Hens in Colony Systems* MAFF, London

FAWC (1993) *Report on Priorities for Animal Welfare Research and Development* MAFF, London

FAWC (undated) *Welfare and Some of the Issues Involved* Ref. PB2667A MAFF, London

FAWC (1996) *Report on the Welfare of Pigs Kept utdoors* MAFF, London

FAWC (1997) *Report on the Welfare of Dairy Cattle* MAFF, London

FAWC (1998) *Codes of Recommendation for the Welfare of Livestock* MAFF, London

FAWC (Dec. 1998) *Report on the Implications of Cloning for the Welfare of Farmed Livestock* MAFF, London

FAWC (1999) *Annual Review 1998* MAFF, London

Ferris C., (1998) Making progress using high genetic merit dairy cows *Grass Farmer* **61** Winter 1998, British Grasslands Society, Reading

Fiddel, N. (1991) *Meat - a Natural Symbol* Routledge, London

Fisher M., (1997) The Role of Ethics in Agriculture *Agricultural Science* 10(4):28-31

Fordham, D., Lincoln, G., Ssewannyana, E., Rodway, R., (1989) Plasma β-endorphin and cortisol concentrations in lambs after handling, transport and slaughter *Animal Production* **49**:103-107

Fraser, A.F., (1954) *Sheep Husbandry* Crosby Lockwood, London

Fraser, A. and Broom, D. (1997) *Farm Animal Behaviour and Welfare* CAB International, Wallingford

Fream, W., (1973) *Elements of Agriculture* John Murray, London

Freedom Food, (1997) *Animal Welfare a Real Issue for Shoppers* RSPCA, Horsham

Freshlay (undated) *The Granary Egg* Freshlay Eggs, Devon

Freudenberger C. D., (1994) What is good agriculture? *Agricultural Ethics* ASA Special Publications **57**

Freudenberger D., (1990) *Global Dust Bowl* Augsburg, Minneapolis

Fussell, G.E., (1969) *The Story of Farming* Pergamon Press, London

Ginever, M. (1997) French Connection *Dairy Farmer* 1st. September 1007

Girling R., (1997) Carnal Desires *The Times* 30th. August 1997

Godfrey J. (April 1998) Ways to alter perceptions *The Pig Industry* British Pig Association Watford

Godfrey J., (1999) *Animal Welfare Suffers as UK Imports More Pigmeat from Illegal Production Units* National Pig Association, London

Grace J., (Nov. 1998) Producers must have their say over antibiotic growth promoters *Pig World*

Gracey, J. (1998) *Meat Plant Operations* Chalcombe Publications, Lincoln

Gunner D., (1989) Market Trends for Meat *Farm Management* 7(2):95-101

Guy R., (1989) Ethical Problems in Farming Practice in *The Status of Animals* (ed. Paterson D., and Palmer M.:87-94) CAB International, Wallingford

Hagedoorn A., (1950) *Animal Breeding* Crosby Lockwood, London

Harrison R., (1964) *Animal Machines* Vincent Stuart Ltd., London

Harrison R., (1987) *Farm Animal Welfare, what, if any, progress?* UFAW, Potters Bar

Hartley M. and Ingilby J., (1980) *The Yorkshire Dales* J.M. Dent and Son Ltd., London

Hartley M. and Ingilby J., (1981) *Life and Tradition in the Yorkshire Dales* Dalesmen Publications Ltd., Skipton

Heffernan, J. (Oct. 1995) Our Daily Bread - the Business of Rural America *Sojourners* Washington, DC

Hemsworth P.H., Barnett J.L. and Coleman G.J. (1993) The Human-animal relationship in agriculture and its consequences for the animal *Animal Welfare* **2**:33-51

Hemsworth P.H., Coleman G.J., (1998) *Human-livestock interactions* CAB International, Wallingford

Henderson D., (1991) *The Veterinary Book for Sheep Farmers* Farming Press, Ipswich

Hill B., (1975) *Britain's Agricultural Industry* Heinemann Educational Books, London

Hiron J, (1996) Pigs fly from little sties *The Agronomist* Spring 1996

HMSO (1947) *Agricultural Act* HMSO, London

HMSO, (1965) *Report of the Technical Enquiry into the Welfare of Animals kept under Intensive Livestock Husbandry Systems* (The Brambell Committee), HMSO, London

House of Commons Agricultural Committee (1981) *Animal Welfare in Poultry, Pig and Veal Calf Production 1980-81* HMSO, London

Howard W., (1989) *Nature's Role in Animal Welfare* UFAW, Potters Bar, Herts

HSA (1996) *Gaining New Ground - HSA Annual Report 1995-96* Humane Slaughter Association, Potters Bar

HSA (1996) *Trustees Report and Accounts 1995-1996* Humane Slaughter Association, Potters Bar

HSA (1998) *'Annual Report 1997-98'* Humane Slaughter Association, Potters Bar

Hume C.W., (1957) *The Status of Animals in the Christian Religion* UFAW, Potters Bar

IFAW (1996) *Ethical Concern for Animals* International Fund for Animal Welfare, Crowborough

Iggo A., (1984) *Pain in Animals* UFAW, Potters Bar

Ingold T, (1994) *From Trust to Domination'* in *'Animals and Human Society* (ed. by Manning and Serpel) Routledge, London

Irwin, A (1996) Duty and the Beasts *The Times Higher* 9th. August 1996

Jenkins, D., (1969) *Living with Questions* SCM Press Ltd., London

Jenkins, L. (1997) Activists sent out guides on DIY anarchy *The Times* 30th. August 1997

Johnson A., (1996) Animals as Food Producers *Food Ethics* (ed. Mepham B.) Routledge, London

Jones C., (1988*) Biblical Signposts for Agricultural Policy* UCCF Publications, Leicester

Kendrick K., (1997) Animal Awareness *Animal Choices* British Society of Animal Science 20:1-8

Kennedy, J., (1992) *The New Anthropomorphism* Cambridge University Press, Cambridge

Kilpatrick J., (1997) Genetically modified organisms – facts and fiction *Agricultural Strategy 1997* ADAS, London

King G., (1998) Industry still waiting *Farming News Spotlight* May 1998

Korten D., (1996) *When Corporations Rule the World* Earthscan Publications, London

Lacey, R., (Feb. 1993) Mad Cows put me off Meat for Life *The Vegetarian*

Lawrence A. and Rushen J., (ed) (1993) *Stereotypic Animal Behaviour – Fundamentals and Applications to Welfare* CAB International, Wallingford

Lawrence A., Pryce J., Stott A., (1999) Animal welfare and public perception as potential constraints to genetic changes in animal production *The Challenge of Genetic Change in Animal Production* British Society of Animal Science, Edinburgh

Leach E., (undated) My calves are going to the dogs *The Independent*

Leahy M., (1991) *Against Liberation - Putting Animals in Perspective* Routledge, London

Leahy M., (1996) Brute Equivocation *The Liberation Debate* (ed. Leahy and Cohn-Sherbok) Routledge, London

Leahy M and Cohn-Sherbok (ed) (1996) *The Liberation Debate* Routledge, London

Leakey R., and Lewin R. (1992) *Origins Reconsidered* Little, Brown and Co., London

Lean I., (1994) Pigs *Management and Welfare of Farm Animals* UFAW, Potters Bar

Leaver J., (1994) Dairy Cattle *Management and Welfare of Farm Animals* UFAW, Potters Bar

Lehman H., (undated) On the Moral Acceptability of Killing Animals *Journal of Agricultural Ethics* 1:155-162

Lewis C., (1957) *The Problem of Pain* Fontana, London

Lewis C., (1979) *God in the Dock* Fount Paperbacks, London

Lewis C., (1989) *First and Second Things* Fount Paperbacks, London

Linzey A., (1976) *Animal Rights* SCM Press, London

Linzey, A. (1994) *Animal Theology* SCM Press Ltd., London

Linzey A., (1998) *Animal Gospel* Hodder and Stoughton, London

Linzey, A. (1996) For Animal Rights *The Liberation Debate* (ed. Leahy and Cohn-Sherbok) Routledge, London

Linzey, A. and Regan, T. (1990) *Animals and Christianity* Crossroad Publishing Co., New York

Linzey, A. and Yamamoto, D. (1998) *Animals on the Agenda* SCM Press Ltd., London

Livingston, J. (1994) *Rogue Primate* Key Porter Books, Toronto, Canada

Lord E. and Whittle D., (1969) *A Theological Glossary* The Religious Education Press, Oxford

MAFF (1990) *Agriculture in the United Kingdom* MAFF Publications, London

MAFF (1996) *Research Strategy 1996-2000* MAFF Publications, London

MAFF (1998) *Consumers can use check-out power to boost welfare standards* MAFF News Release 9th. July 1998, MAFF, London

294

MAFF (1998) *Pig industry needs to sell welfare* MAFF News Release 439/98 11th. November 1998, MAFF, London
MAFF (1998) *Agriculture in the United Kingdom 1997* HMSO, London
MAFF (2000) *Agriculture in the United Kingdom 1999* HMSO, London
Maehle, A. H., (1994) Cruelty and Kindness to the Brute Creation *Animals and Human Society* (ed. Manning A. and Serpel J.) Routledge, London
McDonalds Communication Dept. (July 1998) *'McDonalds: Working with British Agriculture'* McDonalds Communication Department, London
McInerney J., (1991) Assessing the Benefits of Farm Animal Welfare in *Farm Animals - It Pays to be Humane* (ed. Carruthers S.:15-31) Centre for Agricultural Strategy, University of Reading
MacIntyre A., (1985) *Virtue – A Study in Moral Theory* Duckworth, London
Mason G., and Mendl M., (1993) Why is there no simple way of measuring animal welfare? *Animal Welfare .2* :4
Mepham B. (ed.) (1996) *Food Ethics* Routledge, London
Mepham B., (1999) Ethical Impacts of Biotechnology in Dairying in *Progress in Dairy Science* (ed.Phillips C.:388-390) CAB International, Wallingford
Mepham B., Tucker G., and Wiseman J (ed.) (1995) *Issues in Agricultural Bioethics* Nottingham University Press, Nottingham
Methodist Conference Agenda, The (1997) The Methodist Publishing House, Peterborough
Midgley, M. (1986) *Conflicts and Inconsistencies over Animal Welfare* UFAW, Potters Bar
Midgley, M. (1984) *Animals and Why They Matter* The University of Georgia Press, Athens, Georgia
MLC (1994) What the Experts Say *The Vegetarian Choice* Independence Educational Publishers, Cambridge
MLC (1998) *A Pocketful of Meat Facts* Meat and Livestock Commission Economics Service Dept., Milton Keynes
MLC (1999) *A Pocketful of Meat Facts* Meat and Livestock Commission Economics Service Dept., Milton Keynes
Muddiman, J. (1998) A New Testament Doctrine of Creation *Animals on the Agenda* (ed. Linzey, A., and Yamamoto, D.) SCM Press Ltd., London
Murdoch I., (1970) *The Sovereignty of Good* Routledge, London
Naidoo J. and Wills J., (1994) *Health Promotion* Bailliere Tindall, London
National Pig Association (December 1999) *The British Pig Industry Crisis* British Pig Association, London
Newby H., (1988) *The Countryside in Question* Hutchinson, London

NFU (undated) *Building a British Kite Mark* NFU, London

NFU (undated) *NFU Countryside Folder* NFU Countryside, London

NFU, (1988) *The Image of British Farmers in 1988* NFU Publications, London

NFU, (1995) *Caring for Livestock - Report of the NFU Animal Welfare Working Group* NFU Publications, London

NFU, (July 1996) *The BSE Crisis: A Summary of the Current Situation* NFU, London

NFU (1998) *Farming Economy 1998 – Is UK Agriculture Competitive?* NFU Publications, London

NFU (April 1998) *Comments on the critical elements of the EC Commission proposal for a Council directive laying down minimum standards for the protection of laying hens* NFU, London

NFU (Sept. 1999) *Audit for Action – The True Cost of the Farming Crisis* NFU, London

NFU (July 2000) *Poultry Bulletin: Egg Production Quarterly* NFU, Newmarket

NFU (Sept. 2000) *British Milk – What Price?* NFU, London

NFU Media Release (28.3.00) *Farmers offer contract with society at summit* NFU, London

Nicol C., (1997) Environmental Choices of Farm Animals *Animal Choices* British Society of Animal Science **20:**35-43

North J., (1990) *Technology* in *Agriculture in Britain: changing pressures and policies* (ed. Britton D.) CAB International, Wallingford

NPA (8.11.99) *Animal welfare suffers as UK imports more pigmeat from illegal production unit* National Pig Association, London

NPA (28.1.00) *The British pig industry crisis – the economic and social costs* National Pig Association, London

Page, R. (1996) *God and the Web of Creation* SCM Press Ltd., London

Patterson D and Palmer M., (ed.) (1989) *The Status of Animals* CAB International, Wallingford

Pierce A., (1996) Bitter Replay of the Last Big Food Scare *The Times* 23rd. March 1996

Potter M., (March 1995) Time to Play the Winning Welfare Card *The British Farmer* NFU, London

Prescott NB, Mottram TT and Webster AJF, (1997) Relative motivations of dairy cows to attend a voluntary automatic milking system *British Society of Animal Science Occasional Publication* **20:**80-83

Pretty J., (1998) *The Living Land* Earthscan Publications, London

Rackham O., (1995) *The History of the Countryside* Weidenfeld and Nicolson, London

Ravel A., D'Allaire S., Bigras-Poulin M., Ward, R. (1996) Personality traits of stockpeople working in farrowing units on two types of farm in

296

Quebec *Proceedings of the 14th. International Pig Veterinary Society Congress* Bologna, Italy 7th-10th. July 1996:514

Regan T., (1988) *The Case for Animal Rights* Routledge, London

Reiss M., (1998) Building Animals to Order *The Biologist* **45** (4):161-163

Richards D. and Hunt J.W., (1965) *Modern Britain 1783-1964* Longmans, London

Richardson D., (26.5.00) *The Farmers Weekly* Reed Business Information, Sutton

Rifkin, J. (1992) *Beyond Beef* Dutton, London

Ritvo H., (1987) *The Animal Estate* Harvard University Press, Cambridge, Massachusetts

Ritvo, H., (1994) Animals in Nineteenth Century Britain *Animals and Human Society* (ed. Manning A. and Serpell J.) Routledge, London

Rivers J., (1997) Beyond Rights: the morality of rights language *Cambridge Papers* **6** (3)

Rollin B., (1993) *Animal Production, the Beef Industry and the New Social Ethic for Animals* Colorado State University, Colorado

Rolston H., (1988) *Environmental Ethics* Temple University Press, Philadelphia

RSPCA (undated) *Freedom Food – leading the way in farm animal welfare* RSPCA Horsham

RSPCA (undated) *Good News for Farmers* RSPCA, Horsham

RSPCA (1994) *Policies of Animal Welfare* RSPCA, Horsham

RSPCA (1995) *Farm Animal Welfare* RSPCA, Horsham

RSPCA (1995) *The Reverend Arthur Broome, MA (1780-1837)* RSPCA, Horsham

RSPCA (1997) *Focus* **6** RSPCA, Horsham

RSPCA (1997) *Animal Welfare - A real Issue for Shoppers* RSPCA, Horsham

Russell, A (1983) *The Changing Farm: A report of an enquiry into the Ethics of Modern Farming* The Arthur Rank Centre, Kenilworth

Russell B., (1969) *History of Western Philosophy* Allen and Unwin Ltd., London

Russell K. (1963) *Farming with Ken Russell* Farming Press, Ipswich

Russell K., (1969) *The Herdsman's Book* Farming Press, Ipswich

Russell K., (1969) *The Principles of Dairy Farming* Farming Press, Ipswich

Sainsbury D., (1986) *Farm Animal Welfare* Collins, London

Sainsbury's *Nature's Finest Beef* J. Sainsbury plc 729/849, London

Sandoe P., Holtug N., Simonsen H., (1996) Ethical Limits to Domestication *Journal of Agricultural and Environmental Ethics* **9**(2):114-122

Sandoe P., and Simonsen H., Assessing Animal Welfare: where does science end and philosophy begin? *Animal Welfare* **1** (4):257-266

Schwabe C.W., (1994) Animals in the Ancient World *Animals and Human Society* (ed. Manning and Serpel) Routledge, London

Scott, D. (Sept. 1998) The Survey - a question of responsibility *Pig World*

Scott, D. (Jan. 1999) Supermarket Supertankers Start to Turn *Pig World*

Scruton R., (1994) *Modern Philosophy - an introduction and survey* Mandarin Paperbacks, London

Scruton R., (1996) *Animal Rights and Wrongs* Demos, London

Seabrook M, (1984) The psychological interaction between the stockman and his animals and its influence on performance of pigs and dairy cows *The Veterinary Record* **115**:84-87

Seabrook M., (1988) *Stockmanship in Pig Production* RASE, Stoneleigh

Seabrook M., (1991) The human factor - the benefits of humane and skilled stockmanship *It Pays to be Humane* (ed. Carruthers S.) Centre for Agricultural Strategy, University of Reading, UK

Seabrook M., (1994) The role of vernacular knowledge in farming *The Journal of Agricultural Manpower Society* 1 (26):1-9

Seabrook, M. (1996) Stockmanship - Folklore, Fact, Fiction or Fantasy? *The Royal Association of British Dairy Farmers - Dairy Farming Conference* RABDF, Leamington Spa

Seabrook M, and Mount NC, (1995) Individuality in the reaction of pigs to humans *The Pig Journal 1995*

Seabrook M. and Wilkinson M., (1998) *The Stockperson as a Resource on Dairy Farms* The Milk Development Council, London

Seebohm M.E., (1976) *The Evolution of the English Farm* EP Publishing, Wakefield

Serpell J., and Paul E. (1994) Pets and the development of Political Attitudes to Animals *Animals and Human Society* (ed. Manning A. and Serpel J.) Routledge, London

Serpel J., (1989) Attitudes to Animals *The Status of Animals* (ed. Patterson D. and Palmer M.) CAB International, Wallingford, Oxon

Singer P., (1979) *Practical Ethics* Cambridge University Press, Cambridge

Singer P., (1995) *Animal Liberation* Pimlico, London

Soffe R. (1995) (ed.) *Primrose McConnell's The Agricultural Notebook* Blackwell Science, Oxford

Soil Association, The, and The Food Commission (April-June 1995) Factory Farming *Living Earth and The Food Magazine*

Sorrell R.., (1988) *'St. Francis of Assis and Nature'* Oxford University Press, New York

Spedding, C., (1995) Sustainability in Animal Production Systems *Animal Science* **61**:1-8

Spedding C., (2000) *Animal Welfare* Earthscan Publications, London

Society, Religion and Technology (1999) *Is it Right to Clone Animals?* Church of Scotland, Edinburgh

298

Stark B., Machin D., Wilkinson J., (1990) *Outdoor Pigs - Principles and Practice* Chalcombe Press, Lincoln

Stevenson P., (1994) *For Their Own Good* Compassion in World Farming, Petersfield

Stevenson P., (1997) *Factory Farming and the Myth of Cheap Food* Compassion in World Farming, Petersfield

Stewart M., (1989) Teaching of Animal Welfare to Veterinary Students *The Status of Animals* (ed. Paterson D. and Palmer M.) CAB International, Wallingford

Street A.J. (1932) *Farmer's Glory* Faber & Faber, London

Sykes A., (1994) *Laying hens'* in *'Management and Welfare of Farm Animals* UFAW, Potters Bar

Tesco (undated) *Tesco and farmers working together* Tesco Stores Ltd., Cheshunt

Thirsk J., (1957) *English Peasant Farming* Routledge and Kegan Paul, London

Thomas K., (1984) *Man and the Natural World* Penguin Books, London

Toffler, A. (1970) *Future Shock* Pan Books Ltd., London

Tudge C, (1997) The Freaks of the Farmyard *The Times Weekend* 30th. August 1997

Tudge C., (1997) Rights and Wrongs *The Independent on Sunday* 16th. March 1997

UFAW (1993) *Extensive and Organic Livestock Systems - Animal Welfare Implications* UFAW, Potters Bar

UFAW (Dec. 1995) *Animal Welfare* UFAW, Potters Bar.

UFAW (Spring 1996) *UFAW Animal Welfare Research Projects* UFAW, Potters Bar

Vardy P and Grosch P., (1994) *The Puzzle of Ethics* Fount, London

Varley (undated) *The Implications of Welfare on Pig Profitability* Leeds University, Leeds

Varley M., (1991) Stress and Reproduction *Pig News and Information* **12**(4):567-571

Vidal J (2000) Farmers to the Slaughter *The Guardian Weekly* 9th. March 2000

Von Rad G., (1956) *Genesis* SCM Press Ltd., London

Walton, S. (Sept. 1998) Mission Accomplished *Pig World*

Walton S., (Dec. 1999) They Simply Shine *Pig World*

Walton S., (March 2000) Let Health Come First *Pig World*

Warnock M., (1998) *An Intelligent Person's Guide to Ethics* Duckworth, London

Watson J.A.S. and Hobbs M.E., (1951) *Great Farmers* Faber and Faber, London

Watson-Smyth K., (1999) 350 Sheep Dumped on Charity *The Independent* 17th. August 1999

Webster A., (1994) Beef Cattle and Veal Calves *Management and Welfare of Farm Animals* UFAW, Potters Bar

Webster J., (1989) Animal housing as perceived by the animal *Veterinary Annual* **29**:1-7

Webster J., (1990) Outdoor Pig Production - animal welfare and future trends *Outdoor Pigs - Principles and Practice* (ed. Stark *et al*) Chalcombe Press, Lincoln

Webster J., (1991) Farm Animal Welfare, Science and Humanity *The Biologist* **38**(5)

Webster J., (1994) *Animal Welfare, a cool eye towards Eden* Blackwell Science, Oxford

Webster J., (1997) *Animals and Husbandry* RASE, Stoneleigh

Winter M., Fry C., Carruther S., (1998) European Agricultural Policy and Farm Animal Welfare *Food Policy* **23**:3/4:305-323

Wrathall J., (April 1996) Biotechnology - the future is here *The British Farmer* NFU, London

Wright A., (25.9.99) Consumer must be involved *The Farmers Weekly* 25th. September 1999

Wrighton, B (1987) *Reason, Religion and the Animals* Catholic Study Circle for Animal Welfare, London

Young A., (1971) *General view of agriculture in the County of Hertfordshire 1804* David and Charles, Newton Abbot

Young A., (1973) *The Farmer's Kalendar* EP Publishing, Wakefield

Index